James Mills Peirce

# Mathematical Tables

Chiefly to four figures. First Series

James Mills Peirce

**Mathematical Tables**
*Chiefly to four figures. First Series*

ISBN/EAN: 9783337423926

Printed in Europe, USA, Canada, Australia, Japan

Cover: Foto ©berggeist007 / pixelio.de

More available books at **www.hansebooks.com**

# MATHEMATICAL TABLES

## CHIEFLY TO FOUR FIGURES

## FIRST SERIES

BY

## JAMES MILLS PEIRCE

UNIVERSITY PROFESSOR OF MATHEMATICS IN HARVARD
UNIVERSITY

———

BOSTON, U.S.A.:
PUBLISHED BY GINN & COMPANY.
1889.

# CONTENTS.

——◆——

| N | 0 | 1 | 2 | 3 | 4 | 5 | 6 | 7 | 8 | 9 | P. P. 1· 2· 3· 4· 5 |
|---|---|---|---|---|---|---|---|---|---|---|---|
| **10** | 0000 | 0043 | 0086 | 0128 | 0170 | 0212 | 0253 | 0294 | 0334 | 0374 | 4· 8·12·17·21 |
| 11 | 0414 | 0453 | 0492 | 0531 | 0569 | 0607 | 0645 | 0682 | 0719 | 0755 | 4 8·11·15·19 |
| 12 | 0792 | 0828 | 0864 | 0899 | 0934 | 0969 | 1004 | 1038 | 1072 | 1106 | 3· 7·10·14·17 |
| 13 | 1139 | 1173 | 1206 | 1239 | 1271 | 1303 | 1335 | 1367 | 1399 | 1430 | 3· 6·10·13·16 |
| 14 | 1461 | 1492 | 1523 | 1553 | 1584 | 1614 | 1644 | 1673 | 1703 | 1732 | 3· 6· 9·12·15 |
| **15** | 1761 | 1790 | 1818 | 1847 | 1875 | 1903 | 1931 | 1959 | 1987 | 2014 | 3· 6· 8·11·14 |
| 16 | 2041 | 2068 | 2095 | 2122 | 2148 | 2175 | 2201 | 2227 | 2253 | 2279 | 3· 5· 8·11·13 |
| 17 | 2304 | 2330 | 2355 | 2380 | 2405 | 2430 | 2455 | 2480 | 2504 | 2529 | 2· 5· 7·10·12 |
| 18 | 2553 | 2577 | 2601 | 2625 | 2648 | 2672 | 2695 | 2718 | 2742 | 2765 | 2· 5· 7· 9·12 |
| 19 | 2788 | 2810 | 2833 | 2856 | 2878 | 2900 | 2923 | 2945 | 2967 | 2989 | 2· 4· 7· 9·11 |
| **20** | 3010 | 3032 | 3054 | 3075 | 3096 | 3118 | 3139 | 3160 | 3181 | 3201 | 2· 4· 6· 8·11 |
| 21 | 3222 | 3243 | 3263 | 3284 | 3304 | 3324 | 3345 | 3365 | 3385 | 3404 | 2· 4· 6· 8·10 |
| 22 | 3424 | 3444 | 3464 | 3483 | 3502 | 3522 | 3541 | 3560 | 3579 | 3598 | 2· 4· 6· 8·10 |
| 23 | 3617 | 3636 | 3655 | 3674 | 3692 | 3711 | 3729 | 3747 | 3766 | 3784 | 2· 4· 5· 7· 9 |
| 24 | 3802 | 3820 | 3838 | 3856 | 3874 | 3892 | 3909 | 3927 | 3945 | 3962 | 2· 4· 5· 7· 9 |
| **25** | 3979 | 3997 | 4014 | 4031 | 4048 | 4065 | 4082 | 4099 | 4116 | 4133 | 2· 3· 5· 7· 9 |
| 26 | 4150 | 4166 | 4183 | 4200 | 4216 | 4232 | 4249 | 4265 | 4281 | 4298 | 2· 3· 5· 7· 8 |
| 27 | 4314 | 4330 | 4346 | 4362 | 4378 | 4393 | 4409 | 4425 | 4440 | 4456 | 2· 3· 5· 6· 8 |
| 28 | 4472 | 4487 | 4502 | 4518 | 4533 | 4548 | 4564 | 4579 | 4594 | 4609 | 2· 3· 5· 6· 8 |
| 29 | 4624 | 4639 | 4654 | 4669 | 4683 | 4698 | 4713 | 4728 | 4742 | 4757 | 1· 3· 4· 6· 7 |
| **30** | 4771 | 4786 | 4800 | 4814 | 4829 | 4843 | 4857 | 4871 | 4886 | 4900 | 1· 3· 4· 6· 7 |
| 31 | 4914 | 4928 | 4942 | 4955 | 4969 | 4983 | 4997 | 5011 | 5024 | 5038 | 1· 3· 4· 6· 7 |
| 32 | 5051 | 5065 | 5079 | 5092 | 5105 | 5119 | 5132 | 5145 | 5159 | 5172 | 1· 3· 4· 5· 7 |
| 33 | 5185 | 5198 | 5211 | 5224 | 5237 | 5250 | 5263 | 5276 | 5289 | 5302 | 1· 3· 4· 5· 6 |
| 34 | 5315 | 5328 | 5340 | 5353 | 5366 | 5378 | 5391 | 5403 | 5416 | 5428 | 1· 3· 4· 5· 6 |
| **35** | 5441 | 5453 | 5465 | 5478 | 5490 | 5502 | 5514 | 5527 | 5539 | 5551 | 1· 2· 4· 5· 6 |
| 36 | 5563 | 5575 | 5587 | 5599 | 5611 | 5623 | 5635 | 5647 | 5658 | 5670 | 1· 2· 4· 5· 6 |
| 37 | 5682 | 5694 | 5705 | 5717 | 5729 | 5740 | 5752 | 5763 | 5775 | 5786 | 1· 2· 3· 5· 6 |
| 38 | 5798 | 5809 | 5821 | 5832 | 5843 | 5855 | 5866 | 5877 | 5888 | 5899 | 1· 2· 3· 5· 6 |
| 39 | 5911 | 5922 | 5933 | 5944 | 5955 | 5966 | 5977 | 5988 | 5999 | 6010 | 1· 2· 3· 4· 6 |
| **40** | 6021 | 6031 | 6042 | 6053 | 6064 | 6075 | 6085 | 6096 | 6107 | 6117 | 1· 2· 3· 4· 5 |
| 41 | 6128 | 6138 | 6149 | 6160 | 6170 | 6180 | 6191 | 6201 | 6212 | 6222 | 1· 2· 3· 4· 5 |
| 42 | 6232 | 6243 | 6253 | 6263 | 6274 | 6284 | 6294 | 6304 | 6314 | 6325 | 1· 2· 3· 4· 5 |
| 43 | 6335 | 6345 | 6355 | 6365 | 6375 | 6385 | 6395 | 6405 | 6415 | 6425 | 1· 2· 3· 4· 5 |
| 44 | 6435 | 6444 | 6454 | 6464 | 6474 | 6484 | 6493 | 6503 | 6513 | 6522 | 1· 2· 3· 4· 5 |
| **45** | 6532 | 6542 | 6551 | 6561 | 6571 | 6580 | 6590 | 6599 | 6609 | 6618 | 1· 2· 3· 4· 5 |
| 46 | 6628 | 6637 | 6646 | 6656 | 6665 | 6675 | 6684 | 6693 | 6702 | 6712 | 1· 2· 3· 4· 5 |
| 47 | 6721 | 6730 | 6739 | 6749 | 6758 | 6767 | 6776 | 6785 | 6794 | 6803 | 1· 2· 3· 4· 5 |
| 48 | 6812 | 6821 | 6830 | 6839 | 6848 | 6857 | 6866 | 6875 | 6884 | 6893 | 1· 2· 3· 4· 4 |
| 49 | 6902 | 6911 | 6920 | 6928 | 6937 | 6946 | 6955 | 6964 | 6972 | 6981 | 1· 2· 3· 4· 4 |
| **50** | 6990 | 6998 | 7007 | 7016 | 7024 | 7033 | 7042 | 7050 | 7059 | 7067 | 1· 2· 3· 3· 4 |
| 51 | 7076 | 7084 | 7093 | 7101 | 7110 | 7118 | 7126 | 7135 | 7143 | 7152 | 1· 2· 3· 3· 4 |
| 52 | 7160 | 7168 | 7177 | 7185 | 7193 | 7202 | 7210 | 7218 | 7226 | 7235 | 1· 2· 2· 3· 4 |

# Logarithms.

| N | 0 | 1 | 2 | 3 | 4 | 5 | 6 | 7 | 8 | 9 | P. P. 1. 2. 3. 4. 5 |
|---|---|---|---|---|---|---|---|---|---|---|---|
| **55** | 7404 | 7412 | 7419 | 7427 | 7435 | 7443 | 7451 | 7459 | 7466 | 7474 | 1. 2. 2. 3. 4 |
| 56 | 7482 | 7490 | 7497 | 7505 | 7513 | 7520 | 7528 | 7536 | 7543 | 7551 | 1. 2. 2. 3. 4 |
| 57 | 7559 | 7566 | 7574 | 7582 | 7589 | 7597 | 7604 | 7612 | 7619 | 7627 | 1. 2. 2. 3. 4 |
| 58 | 7634 | 7642 | 7649 | 7657 | 7664 | 7672 | 7679 | 7686 | 7694 | 7701 | 1. 1. 2. 3. 4 |
| 59 | 7709 | 7716 | 7723 | 7731 | 7738 | 7745 | 7752 | 7760 | 7767 | 7774 | 1. 1. 2. 3. 4 |
| **60** | 7782 | 7789 | 7796 | 7803 | 7810 | 7818 | 7825 | 7832 | 7839 | 7846 | 1. 1. 2. 3. 4 |
| 61 | 7853 | 7860 | 7868 | 7875 | 7882 | 7889 | 7896 | 7903 | 7910 | 7917 | 1. 1. 2. 3. 4 |
| 62 | 7924 | 7931 | 7938 | 7945 | 7952 | 7959 | 7966 | 7973 | 7980 | 7987 | 1. 1. 2. 3. 3 |
| 63 | 7993 | 8000 | 8007 | 8014 | 8021 | 8028 | 8035 | 8041 | 8048 | 8055 | 1. 1. 2. 3. 3 |
| 64 | 8062 | 8069 | 8075 | 8082 | 8089 | 8096 | 8102 | 8109 | 8116 | 8122 | 1. 1. 2. 3. 3 |
| **65** | 8129 | 8136 | 8142 | 8149 | 8156 | 8162 | 8169 | 8176 | 8182 | 8189 | 1. 1 2. 3. 3 |
| 66 | 8195 | 8202 | 8209 | 8215 | 8222 | 8228 | 8235 | 8241 | 8248 | 8254 | 1. 1. 2. 3. 3 |
| 67 | 8261 | 8267 | 8274 | 8280 | 8287 | 8293 | 8299 | 8306 | 8312 | 8319 | 1. 1. 2. 3. 3 |
| 68 | 8325 | 8331 | 8338 | 8344 | 8351 | 8357 | 8363 | 8370 | 8376 | 8382 | 1. 1. 2. 3. 3 |
| 69 | 8388 | 8395 | 8401 | 8407 | 8414 | 8420 | 8426 | 8432 | 8439 | 8445 | 1. 1. 2. 3. 3 |
| **70** | 8451 | 8457 | 8463 | 8470 | 8476 | 8482 | 8488 | 8494 | 8500 | 8506 | 1. 1. 2. 2. 3 |
| 71 | 8513 | 8519 | 8525 | 8531 | 8537 | 8543 | 8549 | 8555 | 8561 | 8567 | 1. 1. 2. 2. 3 |
| 72 | 8573 | 8579 | 8585 | 8591 | 8597 | 8603 | 8609 | 8615 | 8621 | 8627 | 1. 1. 2. 2. 3 |
| 73 | 8633 | 8639 | 8645 | 8651 | 8657 | 8663 | 8669 | 8675 | 8681 | 8686 | 1. 1. 2. 2. 3 |
| 74 | 8692 | 8698 | 8704 | 8710 | 8716 | 8722 | 8727 | 8733 | 8739 | 8745 | 1. 1. 2. 2. 3 |
| **75** | 8751 | 8756 | 8762 | 8768 | 8774 | 8779 | 8785 | 8791 | 8797 | 8802 | 1. 1. 2. 2. 3 |
| 76 | 8808 | 8814 | 8820 | 8825 | 8831 | 8837 | 8842 | 8848 | 8854 | 8859 | 1. 1. 2. 2. 3 |
| 77 | 8865 | 8871 | 8876 | 8882 | 8887 | 8893 | 8899 | 8904 | 8910 | 8915 | 1. 1. 2. 2. 3 |
| 78 | 8921 | 8927 | 8932 | 8938 | 8943 | 8949 | 8954 | 8960 | 8965 | 8971 | 1. 1. 2. 2. 3 |
| 79 | 8976 | 8982 | 8987 | 8993 | 8998 | 9004 | 9009 | 9015 | 9020 | 9025 | 1. 1. 2. 2. 3 |
| **80** | 9031 | 9036 | 9042 | 9047 | 9053 | 9058 | 9063 | 9069 | 9074 | 9079 | 1. 1. 2. 2. 3 |
| 81 | 9085 | 9090 | 9096 | 9101 | 9106 | 9112 | 9117 | 9122 | 9128 | 9133 | 1. 1. 2. 2. 3 |
| 82 | 9138 | 9143 | 9149 | 9154 | 9159 | 9165 | 9170 | 9175 | 9180 | 9186 | 1. 1. 2. 2. 3 |
| 83 | 9191 | 9196 | 9201 | 9206 | 9212 | 9217 | 9222 | 9227 | 9232 | 9238 | 1. 1. 2. 2. 3 |
| 84 | 9243 | 9248 | 9253 | 9258 | 9263 | 9269 | 9274 | 9279 | 9284 | 9289 | 1. 1. 2. 2. 3 |
| **85** | 9294 | 9299 | 9304 | 9309 | 9315 | 9320 | 9325 | 9330 | 9335 | 9340 | 1. 1. 2. 2. 3 |
| 86 | 9345 | 9350 | 9355 | 9360 | 9365 | 9370 | 9375 | 9380 | 9385 | 9390 | 1. 1. 2. 2. 3 |
| 87 | 9395 | 9400 | 9405 | 9410 | 9415 | 9420 | 9425 | 9430 | 9435 | 9440 | 0. 1. 1. 2. 2 |
| 88 | 9445 | 9450 | 9455 | 9460 | 9465 | 9469 | 9474 | 9479 | 9484 | 9489 | 0. 1. 1. 2. 2 |
| 89 | 9494 | 9499 | 9504 | 9509 | 9513 | 9518 | 9523 | 9528 | 9533 | 9538 | 0. 1. 1. 2. 2 |
| **90** | 9542 | 9547 | 9552 | 9557 | 9562 | 9566 | 9571 | 9576 | 9581 | 9586 | 0. 1. 1. 2. 2 |
| 91 | 9590 | 9595 | 9600 | 9605 | 9609 | 9614 | 9619 | 9624 | 9628 | 9633 | 0. 1. 1. 2. 2 |
| 92 | 9638 | 9643 | 9647 | 9652 | 9657 | 9661 | 9666 | 9671 | 9675 | 9680 | 0. 1. 1. 2. 2 |
| 93 | 9685 | 9689 | 9694 | 9699 | 9703 | 9708 | 9713 | 9717 | 9722 | 9727 | 0. 1. 1. 2. 2 |
| 94 | 9731 | 9736 | 9741 | 9745 | 9750 | 9754 | 9759 | 9763 | 9768 | 9773 | 0. 1. 1. 2. 2 |
| **95** | 9777 | 9782 | 9786 | 9791 | 9795 | 9800 | 9805 | 9809 | 9814 | 9818 | 0. 1. 1. 2. 2 |
| 96 | 9823 | 9827 | 9832 | 9836 | 9841 | 9845 | 9850 | 9854 | 9859 | 9863 | 0. 1. 1. 2. 2 |
| 97 | 9868 | 9872 | 9877 | 9881 | 9886 | 9890 | 9894 | 9899 | 9903 | 9908 | 0. 1. 1. 2. 2 |
| 98 | 9912 | 9917 | 9921 | 9926 | 9930 | 9934 | 9939 | 9943 | 9948 | 9952 | 0 1. 1. 2. 2 |

| N | 0 | 1 | 2 | 3 | 4 | 5 | 6 | 7 | 8 | 9 | 10 |
|---|---|---|---|---|---|---|---|---|---|---|----|
| **100** | 0000 | 0004 | 0009 | 0013 | 0017 | 0022 | 0026 | 0030 | 0035 | 0039 | 0043 |
| 101 | 0043 | 0048 | 0052 | 0056 | 0060 | 0065 | 0069 | 0073 | 0077 | 0082 | 0086 |
| 102 | 0086 | 0090 | 0095 | 0099 | 0103 | 0107 | 0111 | 0116 | 0120 | 0124 | 0128 |
| 103 | 0128 | 0133 | 0137 | 0141 | 0145 | 0149 | 0154 | 0158 | 0162 | 0166 | 0170 |
| 104 | 0170 | 0175 | 0179 | 0183 | 0187 | 0191 | 0195 | 0199 | 0204 | 0208 | 0212 |
| **105** | 0212 | 0216 | 0220 | 0224 | 0228 | 0233 | 0237 | 0241 | 0245 | 0249 | 0253 |
| 106 | 0253 | 0257 | 0261 | 0265 | 0269 | 0273 | 0278 | 0282 | 0286 | 0290 | 0294 |
| 107 | 0294 | 0298 | 0302 | 0306 | 0310 | 0314 | 0318 | 0322 | 0326 | 0330 | 0334 |
| 108 | 0334 | 0338 | 0342 | 0346 | 0350 | 0354 | 0358 | 0362 | 0366 | 0370 | 0374 |
| 109 | 0374 | 0378 | 0382 | 0386 | 0390 | 0394 | 0398 | 0402 | 0406 | 0410 | 0414 |
| **110** | 0414 | 0418 | 0422 | 0426 | 0430 | 0434 | 0438 | 0441 | 0445 | 0449 | 0453 |
| 111 | 0453 | 0457 | 0461 | 0465 | 0469 | 0473 | 0477 | 0481 | 0484 | 0488 | 0492 |
| 112 | 0492 | 0496 | 0500 | 0504 | 0508 | 0512 | 0515 | 0519 | 0523 | 0527 | 0531 |
| 113 | 0531 | 0535 | 0538 | 0542 | 0546 | 0550 | 0554 | 0558 | 0561 | 0565 | 0569 |
| 114 | 0569 | 0573 | 0577 | 0580 | 0584 | 0588 | 0592 | 0596 | 0599 | 0603 | 0607 |
| **115** | 0607 | 0611 | 0615 | 0618 | 0622 | 0626 | 0630 | 0633 | 0637 | 0641 | 0645 |
| 116 | 0645 | 0648 | 0652 | 0656 | 0660 | 0663 | 0667 | 0671 | 0674 | 0678 | 0682 |
| 117 | 0682 | 0686 | 0689 | 0693 | 0697 | 0700 | 0704 | 0708 | 0711 | 0715 | 0719 |
| 118 | 0719 | 0722 | 0726 | 0730 | 0734 | 0737 | 0741 | 0745 | 0748 | 0752 | 0755 |
| 119 | 0755 | 0759 | 0763 | 0766 | 0770 | 0774 | 0777 | 0781 | 0785 | 0788 | 0792 |
| **120** | 0792 | 0795 | 0799 | 0803 | 0806 | 0810 | 0813 | 0817 | 0821 | 0824 | 0828 |
| 121 | 0828 | 0831 | 0835 | 0839 | 0842 | 0846 | 0849 | 0853 | 0856 | 0860 | 0864 |
| 122 | 0864 | 0867 | 0871 | 0874 | 0878 | 0881 | 0885 | 0888 | 0892 | 0896 | 0899 |
| 123 | 0899 | 0903 | 0906 | 0910 | 0913 | 0917 | 0920 | 0924 | 0927 | 0931 | 0934 |
| 124 | 0934 | 0938 | 0941 | 0945 | 0948 | 0952 | 0955 | 0959 | 0962 | 0966 | 0969 |
| **125** | 0969 | 0973 | 0976 | 0980 | 0983 | 0986 | 0990 | 0993 | 0997 | 1000 | 1004 |
| 126 | 1004 | 1007 | 1011 | 1014 | 1017 | 1021 | 1024 | 1028 | 1031 | 1035 | 1038 |
| 127 | 1038 | 1041 | 1045 | 1048 | 1052 | 1055 | 1059 | 1062 | 1065 | 1069 | 1072 |
| 128 | 1072 | 1075 | 1079 | 1082 | 1086 | 1089 | 1092 | 1096 | 1099 | 1103 | 1106 |
| 129 | 1106 | 1109 | 1113 | 1116 | 1119 | 1123 | 1126 | 1129 | 1133 | 1136 | 1139 |
| **130** | 1139 | 1143 | 1146 | 1149 | 1153 | 1156 | 1159 | 1163 | 1166 | 1169 | 1173 |
| 131 | 1173 | 1176 | 1179 | 1183 | 1186 | 1189 | 1193 | 1196 | 1199 | 1202 | 1206 |
| 132 | 1206 | 1209 | 1212 | 1216 | 1219 | 1222 | 1225 | 1229 | 1232 | 1235 | 1239 |
| 133 | 1239 | 1242 | 1245 | 1248 | 1252 | 1255 | 1258 | 1261 | 1265 | 1268 | 1271 |
| 134 | 1271 | 1274 | 1278 | 1281 | 1284 | 1287 | 1290 | 1294 | 1297 | 1300 | 1303 |
| **135** | 1303 | 1307 | 1310 | 1313 | 1316 | 1319 | 1323 | 1326 | 1329 | 1332 | 1335 |
| 136 | 1335 | 1339 | 1342 | 1345 | 1348 | 1351 | 1355 | 1358 | 1361 | 1364 | 1367 |
| 137 | 1367 | 1370 | 1374 | 1377 | 1380 | 1383 | 1386 | 1389 | 1392 | 1396 | 1399 |
| 138 | 1399 | 1402 | 1405 | 1408 | 1411 | 1414 | 1418 | 1421 | 1424 | 1427 | 1430 |
| 139 | 1430 | 1433 | 1436 | 1440 | 1443 | 1446 | 1449 | 1452 | 1455 | 1458 | 1461 |
| **140** | 1461 | 1464 | 1467 | 1471 | 1474 | 1477 | 1480 | 1483 | 1486 | 1489 | 1492 |
| 141 | 1492 | 1495 | 1498 | 1501 | 1504 | 1508 | 1511 | 1514 | 1517 | 1520 | 1523 |
| 142 | 1523 | 1526 | 1529 | 1532 | 1535 | 1538 | 1541 | 1544 | 1547 | 1550 | 1553 |
| 143 | 1553 | 1556 | 1559 | 1562 | 1565 | 1569 | 1572 | 1575 | 1578 | 1581 | 1584 |
| 144 | 1584 | 1587 | 1590 | 1593 | 1596 | 1599 | 1602 | 1605 | 1608 | 1611 | 1614 |
| **145** | 1614 | 1617 | 1620 | 1623 | 1626 | 1629 | 1632 | 1635 | 1638 | 1641 | 1644 |
| 146 | 1644 | 1647 | 1649 | 1652 | 1655 | 1658 | 1661 | 1664 | 1667 | 1670 | 1673 |
| 147 | 1673 | 1676 | 1679 | 1682 | 1685 | 1688 | 1691 | 1694 | 1697 | 1700 | 1703 |
| 148 | 1703 | 1706 | 1708 | 1711 | 1714 | 1717 | 1720 | 1723 | 1726 | 1729 | 1732 |
| 149 | 1732 | 1735 | 1738 | 1741 | 1744 | 1746 | 1749 | 1752 | 1755 | 1758 | 1761 |

# Logarithms.

| N | 0 | 1 | 2 | 3 | 4 | 5 | 6 | 7 | 8 | 9 | 10 |
|---|---|---|---|---|---|---|---|---|---|---|----|
| **150** | 1761 | 1764 | 1767 | 1770 | 1772 | 1775 | 1778 | 1781 | 1784 | 1787 | 1790 |
| 151 | 1790 | 1793 | 1796 | 1798 | 1801 | 1804 | 1807 | 1810 | 1813 | 1816 | 1818 |
| 152 | 1818 | 1821 | 1824 | 1827 | 1830 | 1833 | 1836 | 1838 | 1841 | 1844 | 1847 |
| 153 | 1847 | 1850 | 1853 | 1855 | 1858 | 1861 | 1864 | 1867 | 1870 | 1872 | 1875 |
| 154 | 1875 | 1878 | 1881 | 1884 | 1886 | 1889 | 1892 | 1895 | 1898 | 1901 | 1903 |
| **155** | 1903 | 1906 | 1909 | 1912 | 1915 | 1917 | 1920 | 1923 | 1926 | 1928 | 1931 |
| 156 | 1931 | 1934 | 1937 | 1940 | 1942 | 1945 | 1948 | 1951 | 1953 | 1956 | 1959 |
| 157 | 1959 | 1962 | 1965 | 1967 | 1970 | 1973 | 1976 | 1978 | 1981 | 1984 | 1987 |
| 158 | 1987 | 1989 | 1992 | 1995 | 1998 | 2000 | 2003 | 2006 | 2009 | 2011 | 2014 |
| 159 | 2014 | 2017 | 2019 | 2022 | 2025 | 2028 | 2030 | 2033 | 2036 | 2038 | 2041 |
| **160** | 2041 | 2044 | 2047 | 2049 | 2052 | 2055 | 2057 | 2060 | 2063 | 2066 | 2068 |
| 161 | 2068 | 2071 | 2074 | 2076 | 2079 | 2082 | 2084 | 2087 | 2090 | 2092 | 2095 |
| 162 | 2095 | 2098 | 2101 | 2103 | 2106 | 2109 | 2111 | 2114 | 2117 | 2119 | 2122 |
| 163 | 2122 | 2125 | 2127 | 2130 | 2133 | 2135 | 2138 | 2140 | 2143 | 2146 | 2148 |
| 164 | 2148 | 2151 | 2154 | 2156 | 2159 | 2162 | 2164 | 2167 | 2170 | 2172 | 2175 |
| **165** | 2175 | 2177 | 2180 | 2183 | 2185 | 2188 | 2191 | 2193 | 2196 | 2198 | 2201 |
| 166 | 2201 | 2204 | 2206 | 2209 | 2212 | 2214 | 2217 | 2219 | 2222 | 2225 | 2227 |
| 167 | 2227 | 2230 | 2232 | 2235 | 2238 | 2240 | 2243 | 2245 | 2248 | 2251 | 2253 |
| 168 | 2253 | 2256 | 2258 | 2261 | 2263 | 2266 | 2269 | 2271 | 2274 | 2276 | 2279 |
| 169 | 2279 | 2281 | 2284 | 2287 | 2289 | 2292 | 2294 | 2297 | 2299 | 2302 | 2304 |
| **170** | 2304 | 2307 | 2310 | 2312 | 2315 | 2317 | 2320 | 2322 | 2325 | 2327 | 2330 |
| 171 | 2330 | 2333 | 2335 | 2338 | 2340 | 2343 | 2345 | 2348 | 2350 | 2353 | 2355 |
| 172 | 2355 | 2358 | 2360 | 2363 | 2365 | 2368 | 2370 | 2373 | 2375 | 2378 | 2380 |
| 173 | 2380 | 2383 | 2385 | 2388 | 2390 | 2393 | 2395 | 2398 | 2400 | 2403 | 2405 |
| 174 | 2405 | 2408 | 2410 | 2413 | 2415 | 2418 | 2420 | 2423 | 2425 | 2428 | 2430 |
| **175** | 2430 | 2433 | 2435 | 2438 | 2440 | 2443 | 2445 | 2448 | 2450 | 2453 | 2455 |
| 176 | 2455 | 2458 | 2460 | 2463 | 2465 | 2467 | 2470 | 2472 | 2475 | 2477 | 2480 |
| 177 | 2480 | 2482 | 2485 | 2487 | 2490 | 2492 | 2494 | 2497 | 2499 | 2502 | 2504 |
| 178 | 2504 | 2507 | 2509 | 2512 | 2514 | 2516 | 2519 | 2521 | 2524 | 2526 | 2529 |
| 179 | 2529 | 2531 | 2533 | 2536 | 2538 | 2541 | 2543 | 2545 | 2548 | 2550 | 2553 |
| **180** | 2553 | 2555 | 2558 | 2560 | 2562 | 2565 | 2567 | 2570 | 2572 | 2574 | 2577 |
| 181 | 2577 | 2579 | 2582 | 2584 | 2586 | 2589 | 2591 | 2594 | 2596 | 2598 | 2601 |
| 182 | 2601 | 2603 | 2605 | 2608 | 2610 | 2613 | 2615 | 2617 | 2620 | 2622 | 2625 |
| 183 | 2625 | 2627 | 2629 | 2632 | 2634 | 2636 | 2639 | 2641 | 2643 | 2646 | 2648 |
| 184 | 2648 | 2651 | 2653 | 2655 | 2658 | 2660 | 2662 | 2665 | 2667 | 2669 | 2672 |
| **185** | 2672 | 2674 | 2676 | 2679 | 2681 | 2683 | 2686 | 2688 | 2690 | 2693 | 2695 |
| 186 | 2695 | 2697 | 2700 | 2702 | 2704 | 2707 | 2709 | 2711 | 2714 | 2716 | 2718 |
| 187 | 2718 | 2721 | 2723 | 2725 | 2728 | 2730 | 2732 | 2735 | 2737 | 2739 | 2742 |
| 188 | 2742 | 2744 | 2746 | 2749 | 2751 | 2753 | 2755 | 2758 | 2760 | 2762 | 2765 |
| 189 | 2765 | 2767 | 2769 | 2772 | 2774 | 2776 | 2778 | 2781 | 2783 | 2785 | 2788 |
| **190** | 2788 | 2790 | 2792 | 2794 | 2797 | 2799 | 2801 | 2804 | 2806 | 2808 | 2810 |
| 191 | 2810 | 2813 | 2815 | 2817 | 2819 | 2822 | 2824 | 2826 | 2828 | 2831 | 2833 |
| 192 | 2833 | 2835 | 2838 | 2840 | 2842 | 2844 | 2847 | 2849 | 2851 | 2853 | 2856 |
| 193 | 2856 | 2858 | 2860 | 2862 | 2865 | 2867 | 2869 | 2871 | 2874 | 2876 | 2878 |
| 194 | 2878 | 2880 | 2882 | 2885 | 2887 | 2889 | 2891 | 2894 | 2896 | 2898 | 2900 |
| **195** | 2900 | 2903 | 2905 | 2907 | 2909 | 2911 | 2914 | 2916 | 2918 | 2920 | 2923 |
| 196 | 2923 | 2925 | 2927 | 2929 | 2931 | 2934 | 2936 | 2938 | 2940 | 2942 | 2945 |
| 197 | 2945 | 2947 | 2949 | 2951 | 2953 | 2956 | 2958 | 2960 | 2962 | 2964 | 2967 |
| 198 | 2967 | 2969 | 2971 | 2973 | 2975 | 2978 | 2980 | 2982 | 2984 | 2986 | 2989 |
| 199 | 2989 | 2991 | 2993 | 2995 | 2997 | 2999 | 3002 | 3004 | 3006 | 3008 | 3010 |

| Λ | 6. | 7. | 8. | 9. | 0. | 1. | 2. | 3. |
|---|---|---|---|---|---|---|---|---|
| 00 | 0.0000 | 0.0004 | 0.0043 | 0.0414 9 | 0.3010 50 | 1.0414 91 | 2.0043 | 3.0004 |
| 01 | 0.0000 | 0.0004 | 0.0044 | 0.0423 9 | 0.3061 51 | 1.0505 91 | 2.0142 | 3.0104 |
| 02 | 0.0000 | 0.0005 | 0.0045 | 0.0432 9 | 0.3111 51 | 1.0596 91 | 2.0241 | 3.0204 |
| 03 | 0.0000 | 0.0005 | 0.0046 | 0.0442 10 | 0.3163 52 | 1.0687 91 | 2.0340 | 3.0304 |
| 04 | 0.0000 | 0.0005 | 0.0047 | 0.0452 10 | 0.3215 52 | 1.0779 92 | 2.0439 | 3.0404 |
| 05 | 0.0000 | 0.0005 | 0.0048 | 0.0462 10 | 0.3267 53 | 1.0871 92 | 2.0539 | 3.0504 |
| 06 | 0.0000 | 0.0005 | 0.0050 | 0.0472 10 | 0.3321 53 | 1.0963 92 | 2.0638 | 3.0604 |
| 07 | 0.0001 | 0.0005 | 0.0051 | 0.0482 11 | 0.3374 54 | 1.1055 92 | 2.0737 | 3.0704 |
| 08 | 0.0001 | 0.0005 | 0.0052 | 0.0493 11 | 0.3429 55 | 1.1147 92 | 2.0836 | 3.0804 |
| 09 | 0.0001 | 0.0005 | 0.0053 | 0.0504 11 | 0.3484 55 | 1.1239 92 | 2.0935 | 3.0904 |
| 10 | 0.0001 | 0.0005 | 0.0054 | 0.0515 11 | 0.3539 56 | 1.1332 93 | 2.1034 | 3.1003 |
| 11 | 0.0001 | 0.0006 | 0.0056 | 0.0526 11 | 0.3595 56 | 1.1425 93 | 2.1134 | 3.1103 |
| 12 | 0.0001 | 0.0006 | 0.0057 | 0.0538 12 | 0.3652 57 | 1.1518 93 | 2.1233 | 3.1203 |
| 13 | 0.0001 | 0.0006 | 0.0058 | 0.0550 12 | 0.3709 57 | 1.1611 93 | 2.1332 | 3.1303 |
| 14 | 0.0001 | 0.0006 | 0.0060 | 0.0562 12 | 0.3766 58 | 1.1704 93 | 2.1431 | 3.1403 |
| 15 | 0.0001 | 0.0006 | 0.0061 | 0.0574 12 | 0.3825 59 | 1.1797 93 | 2.1531 | 3.1503 |
| 16 | 0.0001 | 0.0006 | 0.0062 | 0.0586 13 | 0.3884 59 | 1.1891 94 | 2.1630 | 3.1603 |
| 17 | 0.0001 | 0.0006 | 0.0064 | 0.0599 13 | 0.3943 60 | 1.1984 94 | 2.1729 | 3.1703 |
| 18 | 0.0001 | 0.0007 | 0.0065 | 0.0612 13 | 0.4003 60 | 1.2078 94 | 2.1829 | 3.1803 |
| 19 | 0.0001 | 0.0007 | 0.0067 | 0.0625 13 | 0.4063 61 | 1.2172 94 | 2.1928 | 3.1903 |
| 20 | 0.0001 | 0.0007 | 0.0068 | 0.0639 14 | 0.4124 61 | 1.2266 94 | 2.2027 | 3.2003 |
| 21 | 0.0001 | 0.0007 | 0.0070 | 0.0653 14 | 0.4186 62 | 1.2360 94 | 2.2127 | 3.2103 |
| 22 | 0.0001 | 0.0007 | 0.0071 | 0.0667 14 | 0.4248 62 | 1.2454 94 | 2.2226 | 3.2203 |
| 23 | 0.0001 | 0.0007 | 0.0073 | 0.0681 15 | 0.4311 63 | 1.2548 94 | 2.2325 | 3.2303 |
| 24 | 0.0001 | 0.0008 | 0.0075 | 0.0696 15 | 0.4374 63 | 1.2643 95 | 2.2425 | 3.2402 |
| 25 | 0.0001 | 0.0008 | 0.0077 | 0.0711 15 | 0.4438 64 | 1.2738 95 | 2.2524 | 3.2502 |
| 26 | 0.0001 | 0.0008 | 0.0078 | 0.0726 15 | 0.4502 65 | 1.2832 95 | 2.2624 | 3.2602 |
| 27 | 0.0001 | 0.0008 | 0.0080 | 0.0742 16 | 0.4567 65 | 1.2927 95 | 2.2723 | 3.2702 |
| 28 | 0.0001 | 0.0008 | 0.0082 | 0.0757 16 | 0.4632 66 | 1.3022 95 | 2.2823 | 3.2802 |
| 29 | 0.0001 | 0.0008 | 0.0084 | 0.0774 16 | 0.4698 66 | 1.3117 95 | 2.2922 | 3.2902 |
| 30 | 0.0001 | 0.0009 | 0.0086 | 0.0790 17 | 0.4764 67 | 1.3212 95 | 2.3022 | 3.3002 |
| 31 | 0.0001 | 0.0009 | 0.0088 | 0.0807 17 | 0.4831 67 | 1.3308 95 | 2.3121 | 3.3102 |
| 32 | 0.0001 | 0.0009 | 0.0090 | 0.0824 17 | 0.4899 68 | 1.3403 95 | 2.3221 | 3.3202 |
| 33 | 0.0001 | 0.0009 | 0.0092 | 0.0841 18 | 0.4966 68 | 1.3499 96 | 2.3320 | 3.3302 |
| 34 | 0.0001 | 0.0009 | 0.0094 | 0.0859 18 | 0.5035 69 | 1.3594 96 | 2.3420 | 3.3402 |
| 35 | 0.0001 | 0.0010 | 0.0096 | 0.0877 18 | 0.5104 69 | 1.3690 96 | 2.3519 | 3.3502 |
| 36 | 0.0001 | 0.0010 | 0.0098 | 0.0896 19 | 0.5173 70 | 1.3786 96 | 2.3619 | 3.3602 |
| 37 | 0.0001 | 0.0010 | 0.0101 | 0.0915 19 | 0.5243 70 | 1.3881 96 | 2.3718 | 3.3702 |
| 38 | 0.0001 | 0.0010 | 0.0103 | 0.0934 19 | 0.5313 71 | 1.3977 96 | 2.3818 | 3.3802 |
| 39 | 0.0001 | 0.0011 | 0.0105 | 0.0953 20 | 0.5384 71 | 1.4073 96 | 2.3918 | 3.3902 |
| 40 | 0.0001 | 0.0011 | 0.0108 | 0.0973 20 | 0.5455 72 | 1.4170 96 | 2.4017 | 3.4002 |
| 41 | 0.0001 | 0.0011 | 0.0110 | 0.0993 20 | 0.5527 72 | 1.4266 96 | 2.4117 | 3.4102 |
| 42 | 0.0001 | 0.0011 | 0.0113 | 0.1014 21 | 0.5599 72 | 1.4362 96 | 2.4216 | 3.4202 |
| 43 | 0.0001 | 0.0012 | 0.0115 | 0.1035 21 | 0.5672 73 | 1.4458 96 | 2.4316 | 3.4302 |
| 44 | 0.0001 | 0.0012 | 0.0118 | 0.1057 22 | 0.5745 73 | 1.4555 96 | 2.4416 | 3.4402 |
| 45 | 0.0001 | 0.0012 | 0.0121 | 0.1078 22 | 0.5819 74 | 1.4651 97 | 2.4515 | 3.4502 |
| 46 | 0.0001 | 0.0013 | 0.0123 | 0.1101 22 | 0.5893 74 | 1.4748 97 | 2.4615 | 3.4602 |
| 47 | 0.0001 | 0.0013 | 0.0126 | 0.1123 23 | 0.5967 75 | 1.4845 97 | 2.4715 | 3.4701 |
| 48 | 0.0001 | 0.0013 | 0.0129 | 0.1146 23 | 0.6042 75 | 1.4941 97 | 2.4814 | 3.4801 |
| 49 | 0.0001 | 0.0013 | 0.0132 | 0.1169 24 | 0.6118 76 | 1.5038 97 | 2.4914 | 3.4901 |
| 50 | 0.0001 | 0.0014 | 0.0135 | 0.1193 24 | 0.6193 76 | 1.5135 97 | 2.5014 | 3.5001 |

## Logarithms of Sums and Differences.

| A | 6. | 7. | 8. | 9. | 0. | 1. | 2. | 3. |
|---|---|---|---|---|---|---|---|---|
| **50** | 0.0001 | 0.0014 | 0.0135 | 0.1193 24 | 0.6193 76 | 1.5135 97 | 2.5014 | 3.5001 |
| 51 | 0.0001 | 0.0014 | 0.0138 | 0.1218 24 | 0.6269 76 | 1.5232 97 | 2.5113 | 3.5101 |
| 52 | 0.0001 | 0.0014 | 0.0141 | 0.1242 25 | 0.6346 77 | 1.5329 97 | 2.5213 | 3.5201 |
| 53 | 0.0001 | 0.0015 | 0.0145 | 0.1267 25 | 0.6423 77 | 1.5426 97 | 2.5313 | 3.5301 |
| 54 | 0.0002 | 0.0015 | 0.0148 | 0.1293 26 | 0.6501 78 | 1.5523 97 | 2.5413 | 3.5401 |
| **55** | 0.0002 | 0.0015 | 0.0151 | 0.1319 26 | 0.6578 78 | 1.5621 97 | 2.5512 | 3.5501 |
| 56 | 0.0002 | 0.0016 | 0.0155 | 0.1345 27 | 0.6657 78 | 1.5718 97 | 2.5612 | 3.5601 |
| 57 | 0.0002 | 0.0016 | 0.0158 | 0.1372 27 | 0.6735 79 | 1.5815 97 | 2.5712 | 3.5701 |
| 58 | 0.0002 | 0.0016 | 0.0162 | 0.1399 28 | 0.6814 79 | 1.5913 97 | 2.5811 | 3.5801 |
| 59 | 0.0002 | 0.0017 | 0.0166 | 0.1427 28 | 0.6893 80 | 1.6010 97 | 2.5911 | 3.5901 |
| **60** | 0.0002 | 0.0017 | 0.0170 | 0.1455 28 | 0.6973 80 | 1.6108 98 | 2.6011 | 3.6001 |
| 61 | 0.0002 | 0.0018 | 0.0173 | 0.1484 29 | 0.7053 80 | 1.6205 98 | 2.6111 | 3.6101 |
| 62 | 0.0002 | 0.0018 | 0.0177 | 0.1513 29 | 0.7134 81 | 1.6303 98 | 2.6210 | 3.6201 |
| 63 | 0.0002 | 0.0018 | 0.0181 | 0.1543 30 | 0.7215 81 | 1.6401 98 | 2.6310 | 3.6301 |
| 64 | 0.0002 | 0.0019 | 0.0186 | 0.1573 30 | 0.7296 81 | 1.6498 98 | 2.6410 | 3.6401 |
| **65** | 0.0002 | 0.0019 | 0.0190 | 0.1604 31 | 0.7377 82 | 1.6596 98 | 2.6510 | 3.6501 |
| 66 | 0.0002 | 0.0020 | 0.0194 | 0.1635 31 | 0.7459 82 | 1.6694 98 | 2.6609 | 3.6601 |
| 67 | 0.0002 | 0.0020 | 0.0199 | 0.1666 32 | 0.7541 82 | 1.6792 98 | 2.6709 | 3.6701 |
| 68 | 0.0002 | 0.0021 | 0.0203 | 0.1699 32 | 0.7624 83 | 1.6890 98 | 2.6809 | 3.6801 |
| 69 | 0.0002 | 0.0021 | 0.0208 | 0.1731 33 | 0.7707 83 | 1.6988 98 | 2.6909 | 3.6901 |
| **70** | 0.0002 | 0.0022 | 0.0212 | 0.1764 33 | 0.7790 83 | 1.7086 98 | 2.7009 | 3.7001 |
| 71 | 0.0002 | 0.0022 | 0.0217 | 0.1798 34 | 0.7874 84 | 1.7184 98 | 2.7108 | 3.7101 |
| 72 | 0.0002 | 0.0023 | 0.0222 | 0.1832 34 | 0.7957 84 | 1.7282 98 | 2.7208 | 3.7201 |
| 73 | 0.0002 | 0.0023 | 0.0227 | 0.1867 35 | 0.8042 84 | 1.7380 98 | 2.7308 | 3.7301 |
| 74 | 0.0002 | 0.0024 | 0.0232 | 0.1902 35 | 0.8126 85 | 1.7478 98 | 2.7408 | 3.7401 |
| **75** | 0.0002 | 0.0024 | 0.0238 | 0.1938 36 | 0.8211 85 | 1.7577 98 | 2.7508 | 3.7501 |
| 76 | 0.0002 | 0.0025 | 0.0243 | 0.1974 37 | 0.8296 85 | 1.7675 98 | 2.7608 | 3.7601 |
| 77 | 0.0003 | 0.0025 | 0.0248 | 0.2011 37 | 0.8381 85 | 1.7773 98 | 2.7707 | 3.7701 |
| 78 | 0.0003 | 0.0026 | 0.0254 | 0.2048 38 | 0.8467 86 | 1.7871 98 | 2.7807 | 3.7801 |
| 79 | 0.0003 | 0.0027 | 0.0260 | 0.2086 38 | 0.8553 86 | 1.7970 98 | 2.7907 | 3.7901 |
| **80** | 0.0003 | 0.0027 | 0.0266 | 0.2124 39 | 0.8639 86 | 1.8068 98 | 2.8007 | 3.8001 |
| 81 | 0.0003 | 0.0028 | 0.0272 | 0.2163 39 | 0.8725 87 | 1.8167 98 | 2.8107 | 3.8101 |
| 82 | 0.0003 | 0.0029 | 0.0278 | 0.2203 40 | 0.8812 87 | 1.8265 99 | 2.8207 | 3.8201 |
| 83 | 0.0003 | 0.0029 | 0.0284 | 0.2243 40 | 0.8899 87 | 1.8364 99 | 2.8306 | 3.8301 |
| 84 | 0.0003 | 0.0030 | 0.0291 | 0.2284 41 | 0.8986 87 | 1.8462 99 | 2.8406 | 3.8401 |
| **85** | 0.0003 | 0.0031 | 0.0297 | 0.2325 41 | 0.9074 88 | 1.8561 99 | 2.8506 | 3.8501 |
| 86 | 0.0003 | 0.0031 | 0.0304 | 0.2366 42 | 0.9162 88 | 1.8660 99 | 2.8606 | 3.8601 |
| 87 | 0.0003 | 0.0032 | 0.0311 | 0.2409 43 | 0.9250 88 | 1.8758 99 | 2.8706 | 3.8701 |
| 88 | 0.0003 | 0.0033 | 0.0318 | 0.2452 43 | 0.9338 88 | 1.8857 99 | 2.8806 | 3.8801 |
| 89 | 0.0003 | 0.0034 | 0.0325 | 0.2495 44 | 0.9426 89 | 1.8956 99 | 2.8906 | 3.8901 |
| **90** | 0.0003 | 0.0034 | 0.0332 | 0.2539 44 | 0.9515 89 | 1.9054 99 | 2.9005 | 3.9001 |
| 91 | 0.0004 | 0.0035 | 0.0339 | 0.2584 45 | 0.9604 89 | 1.9153 99 | 2.9105 | 3.9101 |
| 92 | 0.0004 | 0.0036 | 0.0347 | 0.2629 45 | 0.9693 89 | 1.9252 99 | 2.9205 | 3.9201 |
| 93 | 0.0004 | 0.0037 | 0.0355 | 0.2674 46 | 0.9782 89 | 1.9351 99 | 2.9305 | 3.9301 |
| 94 | 0.0004 | 0.0038 | 0.0363 | 0.2721 47 | 0.9872 90 | 1.9450 99 | 2.9405 | 3.9400 |
| **95** | 0.0004 | 0.0039 | 0.0371 | 0.2767 47 | 0.9962 90 | 1.9548 99 | 2.9505 | 3.9500 |
| 96 | 0.0004 | 0.0039 | 0.0379 | 0.2815 48 | 1.0052 90 | 1.9647 99 | 2.9605 | 3.9600 |
| 97 | 0.0004 | 0.0040 | 0.0387 | 0.2863 48 | 1.0142 90 | 1.9746 99 | 2.9705 | 3.9700 |
| 98 | 0.0004 | 0.0041 | 0.0396 | 0.2911 49 | 1.0232 91 | 1.9845 99 | 2.9805 | 3.9800 |
| 99 | 0.0004 | 0.0042 | 0.0405 | 0.2961 49 | 1.0323 91 | 1.9944 99 | 2.9904 | 3.9900 |
| **00** | 0.0004 | 0.0043 | 0.0414 | 0.3010 50 | 1.0414 91 | 2.0043 99 | 3.0004 | 4.0000 |

| φ | 0° l sin l tn | l sc | 1° l sin l tn | l sc | 2° l sin l tn | l sc | |
|---|---|---|---|---|---|---|---|
| 00' | − ∞ | 00 | 8.2419 19 72 | 01 | 8.5428 31 36 | 03 | 60' |
| 01' | 6.4637 37 | 00 | 8.2490 91 71 | 01 | 8.5464 67 36 | 03 | 59' |
| 02' | 6.7648 48 | 00 | 8.2561 62 70 | 01 | 8.5500 03 36 | 03 | 58' |
| 03' | 6.9408 08 | 00 | 8.2630 31 69 | 01 | 8.5535 38 35 | 03 | 57' |
| 04' | 7.0658 58 | 00 | 8.2699*00 68 | 01 | 8.5571 73 35 | 03 | 56' |
| 05' | 7.1627 27 | 00 | 8.2766 67 67 | 01 | 8.5605 08 35 | 03 | 55' |
| 06' | 7.2419 19 | 00 | 8.2832 33 66 | 01 | 8.5640 43 34 | 03 | 54' |
| 07' | 7.3088 88 | 00 | 8.2898 99 65 | 01 | 8.5674 77 34 | 03 | 53' |
| 08' | 7.3668 68 | 00 | 8.2962 63 64 | 01 | 8.5708 11 34 | 03 | 52' |
| 09' | 7.4180 80 | 00 | 8.3025 26 63 | 01 | 8.5742 45 34 | 03 | 51' |
| 10' | 7.4637 37 | 00 | 8.3088 89 62 | 01 | 8.5776 79 33 | 03 | 50' |
| 11' | 7.5051 61 | 00 | 8.3150 50 61 | 01 | 8.5809 12 33 | 03 | 49' |
| 12' | 7.5429 29 | 00 | 8.3210 11 60 | 01 | 8.5842 45 33 | 03 | 48' |
| 13' | 7.5777 77 | 00 | 8.3270 71 59 | 01 | 8.5875 78 33 | 03 | 47' |
| 14' | 7.6099 99 | 00 | 8.3329 30 59 | 01 | 8.5907 11 32 | 03 | 46' |
| 15' | 7.6398 98 | 00 | 8.3388 89 58 | 01 | 8.5939 43 32 | 03 | 45' |
| 16' | 7.6678 78 | 00 | 8.3445 46 57 | 01 | 8.5972 75 32 | 03 | 44' |
| 17' | 7.6942 42 | 00 | 8.3502 03 56 | 01 | 8.6003 07 32 | 03 | 43' |
| 18' | 7.7190 90 | 00 | 8.3558 59 56 | 01 | 8.6035 38 31 | 04 | 42' |
| 19' | 7.7425 25 | 00 | 8.3613 14 55 | 01 | 8.6066 70 31 | 04 | 41' |
| 20' | 7.7648 48 | 00 | 8.3668 69 54 | 01 | 8.6097*01 31 | 04 | 40' |
| 21' | 7.7859 60 | 00 | 8.3722 23 54 | 01 | 8.6128 32 31 | 04 | 39' |
| 22' | 7.8061 62 | 00 | 8.3775 78 53 | 01 | 8.6159 63 31 | 04 | 38' |
| 23' | 7.8255 56 | 00 | 8.3828 29 52 | 01 | 8.6189 93 30 | 04 | 37' |
| 24' | 7.8439 39 | 00 | 8.3880 81 52 | 01 | 8.6220 23 30 | 04 | 36' |
| 25' | 7.8617 17 | 00 | 8.3931 32 51 | 01 | 8.6250 54 30 | 04 | 35' |
| 26' | 7.8787 87 | 00 | 8.3982 83 51 | 01 | 8.6279 83 30 | 04 | 34' |
| 27' | 7.8951 51 | 00 | 8.4032 33 50 | 01 | 8.6309 13 30 | 04 | 33' |
| 28' | 7.9109 09 | 00 | 8.4082 83 49 | 01 | 8.6339 43 29 | 04 | 32' |
| 29' | 7.9261 61 | 00 | 8.4131 32 49 | 01 | 8.6368 72 29 | 04 | 31' |
| 30' | 7.9408 09 | 00 | 8.4179 81 48 | 01 | 8.6397*01 29 | 04 | 30' |
| 31' | 7.9551 51 | 00 | 8.4227 29 48 | 02 | 8.6426 30 29 | 04 | 29' |
| 32' | 7.9689 89 | 00 | 8.4275 76 47 | 02 | 8.6454 59 29 | 04 | 28' |
| 33' | 7.9822 23 132 | 00 | 8.4322 23 47 | 02 | 8.6483 87 28 | 04 | 27' |
| 34' | 7.9952 52 128 | 00 | 8.4368 70 46 | 02 | 8.6511 15 28 | 04 | 26' |
| 35' | 8.0078 78 127 | 00 | 8.4414 16 46 | 02 | 8.6539 44 28 | 04 | 25' |
| 36' | 8.0200 00 121 | 00 | 8.4459 61 45 | 02 | 8.6567 71 28 | 04 | 24' |
| 37' | 8.0319 19 117 | 00 | 8.4504 06 45 | 02 | 8.6595 99 28 | 05 | 23' |
| 38' | 8.0435 35 114 | 00 | 8.4549 51 44 | 02 | 8.6622 27 27 | 05 | 22' |
| 39' | 8.0548 48 111 | 00 | 8.4593 95 44 | 02 | 8.6650 54 27 | 05 | 21' |
| 40' | 8.0658 58 109 | 00 | 8.4637 38 43 | 02 | 8.6677 82 27 | 05 | 20' |
| 41' | 8.0765 65 106 | 00 | 8.4680 82 43 | 02 | 8.6704 09 27 | 05 | 19' |
| 42' | 8.0870 70 103 | 00 | 8.4723 25 43 | 02 | 8.6731 36 27 | 05 | 18' |
| 43' | 8.0972 72 101 | 00 | 8.4765 67 42 | 02 | 8.6758 62 27 | 05 | 17' |
| 44' | 8.1072 72 99 | 00 | 8.4807 09 42 | 02 | 8.6784 89 26 | 05 | 16' |
| 45' | 8.1169 70 97 | 00 | 8.4848 51 41 | 02 | 8.6810 15 26 | 05 | 15' |
| 46' | 8.1265 65 94 | 00 | 8.4890 92 41 | 02 | 8.6837 42 26 | 05 | 14' |
| 47' | 8.1358 59 92 | 00 | 8.4930 33 41 | 02 | 8.6863 68 26 | 05 | 13' |
| 48' | 8.1450 50 90 | 00 | 8.4971 73 40 | 02 | 8.6889 94 26 | 05 | 12' |
| 49' | 8.1539 40 89 | 00 | 8.5011 13 40 | 02 | 8.6914 20 26 | 05 | 11' |
| 50' | 8.1627 27 87 | 00 | 8.5050 53 39 | 02 | 8.6940 45 26 | 05 | 10' |
| 51' | 8.1713 13 85 | 00 | 8.5090 92 39 | 02 | 8.6965 71 25 | 05 | 09' |
| 52' | 8.1797 98 84 | 00 | 8.5129 31 39 | 02 | 8.6991 96 25 | 05 | 08' |
| 53' | 8.1880 80 82 | 01 | 8.5167 70 38 | 02 | 8.7016 21 25 | 06 | 07' |

# Logarithms of Circular Functions.

| φ | 3° l sin   l tn | | l sc | 4° l sin   l tn | | l sc | 5° l sin   l tn | | l sc | |
|---|---|---|---|---|---|---|---|---|---|---|
| 00' | 8.7188 94 | 24 | 06 | 8.8436 46 | 18 | 11 | 8.9403 20 | 14 | 17 | 60' |
| 01' | 8.7212 18 | 24 | 06 | 8.8454 65 | 18 | 11 | 8.9417 34 | 14 | 17 | 59' |
| 02' | 8.7236 42 | 24 | 06 | 8.8472 83 | 18 | 11 | 8.9432 49 | 14 | 17 | 58' |
| 03' | 8.7260 66 | 24 | 06 | 8.8490*01 | 18 | 11 | 8.9446 63 | 14 | 17 | 57' |
| 04' | 8.7283 90 | 24 | 06 | 8.8508 18 | 18 | 11 | 8.9460 77 | 14 | 17 | 56' |
| 05' | 8.7307 13 | 23 | 06 | 8.8525 36 | 18 | 11 | 8.9475 92 | 14 | 17 | 55' |
| 06' | 8.7330 37 | 23 | 06 | 8.8543 54 | 18 | 11 | 8.9489*06 | 14 | 17 | 54' |
| 07' | 8.7354 60 | 23 | 06 | 8.8560 72 | 18 | 11 | 8.9503 20 | 14 | 17 | 53' |
| 08' | 8.7377 83 | 23 | 06 | 8.8578 89 | 18 | 11 | 8.9517 34 | 14 | 17 | 52' |
| 09' | 8.7400 06 | 23 | 07 | 8.8595*07 | 17 | 11 | 8.9531 49 | 14 | 18 | 51' |
| 10' | 8.7423 29 | 23 | 07 | 8.8613 24 | 17 | 11 | 8.9545 63 | 14 | 18 | 50' |
| 11' | 8.7445 52 | 23 | 07 | 8.8630 42 | 17 | 12 | 8.9559 77 | 14 | 18 | 49' |
| 12' | 8.7468 75 | 23 | 07 | 8.8647 59 | 17 | 12 | 8.9573 91 | 14 | 18 | 48' |
| 13' | 8.7491 97 | 23 | 07 | 8.8665 76 | 17 | 12 | 8.9587*05 | 14 | 18 | 47' |
| 14' | 8.7513 20 | 22 | 07 | 8.8682 94 | 17 | 12 | 8.9601 19 | 14 | 18 | 46' |
| 15' | 8.7535 42 | 22 | 07 | 8.8699*11 | 17 | 12 | 8.9614 33 | 14 | 18 | 45' |
| 16' | 8.7557 65 | 22 | 07 | 8.8716 28 | 17 | 12 | 8.9628 46 | 14 | 18 | 44' |
| 17' | 8.7580 87 | 22 | 07 | 8.8733 45 | 17 | 12 | 8.9642 60 | 14 | 18 | 43' |
| 18' | 8.7602 09 | 22 | 07 | 8.8749 62 | 17 | 12 | 8.9655 74 | 14 | 19 | 42' |
| 19' | 8.7623 31 | 22 | 07 | 8.8766 78 | 17 | 12 | 8.9669 88 | 14 | 19 | 41' |
| 20' | 8.7645 52 | 22 | 07 | 8.8783 95 | 17 | 12 | 8.9682*01 | 14 | 19 | 40' |
| 21' | 8.7667 74 | 22 | 07 | 8.8799*12 | 17 | 13 | 8.9696*15 | 14 | 19 | 39' |
| 22' | 8.7688 96 | 22 | 08 | 8.8816 29 | 17 | 13 | 8.9709 29 | 14 | 19 | 38' |
| 23' | 8.7710 17 | 21 | 08 | 8.8833 45 | 17 | 13 | 8.9723 42 | 13 | 19 | 37' |
| 24' | 8.7731 39 | 21 | 08 | 8.8849 62 | 16 | 13 | 8.9736 56 | 13 | 19 | 36' |
| 25' | 8.7752 60 | 21 | 08 | 8.8865 78 | 16 | 13 | 8.9750 69 | 13 | 19 | 35' |
| 26' | 8.7773 81 | 21 | 08 | 8.8882 95 | 16 | 13 | 8.9763 82 | 13 | 20 | 34' |
| 27' | 8.7794*02 | 21 | 08 | 8.8898*11 | 16 | 13 | 8.9776 96 | 13 | 20 | 33' |
| 28' | 8.7815 23 | 21 | 08 | 8.8914 27 | 16 | 13 | 8.9789*09 | 13 | 20 | 32' |
| 29' | 8.7836 44 | 21 | 08 | 8.8930 44 | 16 | 13 | 8.9803 23 | 13 | 20 | 31' |
| 30' | 8.7857 65 | 21 | 08 | 8.8946 60 | 16 | 13 | 8.9816 36 | 13 | 20 | 30' |
| 31' | 8.7877 86 | 21 | 08 | 8.8962 76 | 16 | 14 | 8.9829 49 | 13 | 20 | 29' |
| 32' | 8.7898*06 | 20 | 08 | 8.8978 92 | 16 | 14 | 8.9842 62 | 13 | 20 | 28' |
| 33' | 8.7918 27 | 20 | 08 | 8.8994*08 | 16 | 14 | 8.9855 75 | 13 | 20 | 27' |
| 34' | 8.7939 47 | 20 | 08 | 8.9010 24 | 16 | 14 | 8.9868 88 | 13 | 21 | 26' |
| 35' | 8.7959 67 | 20 | 08 | 8.9026 40 | 16 | 14 | 8.9881*01 | 13 | 21 | 25' |
| 36' | 8.7979 88 | 20 | 09 | 8.9042 56 | 16 | 14 | 8.9894*15 | 13 | 21 | 24' |
| 37' | 8.7999*08 | 20 | 09 | 8.9057 71 | 16 | 14 | 8.9907 28 | 13 | 21 | 23' |
| 38' | 8.8019 28 | 20 | 09 | 8.9073 87 | 16 | 14 | 8.9919 40 | 13 | 21 | 22' |
| 39' | 8.8039 48 | 20 | 09 | 8.9089*03 | 16 | 14 | 8.9932 53 | 13 | 21 | 21' |
| 40' | 8.8059 67 | 20 | 09 | 8.9104 18 | 16 | 14 | 8.9945 66 | 13 | 21 | 20' |
| 41' | 8.8078 87 | 20 | 09 | 8.9119 34 | 15 | 15 | 8.9958 79 | 13 | 21 | 19' |
| 42' | 8.8098*07 | 20 | 09 | 8.9135 50 | 15 | 15 | 8.9970 92 | 13 | 22 | 18' |
| 43' | 8.8117 26 | 19 | 09 | 8.9150 65 | 15 | 15 | 8.9983*05 | 13 | 22 | 17' |
| 44' | 8.8137 46 | 19 | 09 | 8.9166 80 | 15 | 15 | 8.9996*17 | 13 | 22 | 16' |
| 45' | 8.8156 65 | 19 | 09 | 8.9181 96 | 15 | 15 | 9.0008 30 | 13 | 22 | 15' |
| 46' | 8.8175 85 | 19 | 09 | 8.9196*11 | 15 | 15 | 9.0021 43 | 13 | 22 | 14' |
| 47' | 8.8194*04 | 19 | 09 | 8.9211 26 | 15 | 15 | 9.0033 55 | 13 | 22 | 13' |
| 48' | 8.8213 23 | 19 | 10 | 8.9226 41 | 15 | 15 | 9.0046 68 | 12 | 22 | 12' |
| 49' | 8.8232 42 | 19 | 10 | 8.9241 56 | 15 | 16 | 9.0058 80 | 12 | 22 | 11' |
| 50' | 8.8251 61 | 19 | 10 | 8.9256 72 | 15 | 15 | 9.0070 93 | 12 | 23 | 10' |
| 51' | 8.8270 80 | 19 | 10 | 8.9271 87 | 15 | 16 | 9.0083*06 | 12 | 23 | 09' |
| 52' | 8.8289 99 | 19 | 10 | 8.9286*02 | 15 | 16 | 9.0095*18 | 12 | 23 | 08' |
| 53' | 8.8307 17 | 19 | 10 | 8.9301 16 | 15 | 16 | 9.0107 30 | 12 | 23 | 07' |

# Logarithms of Circular Functions.

| φ | S | l sin φ | φ | T | l tan φ | φ | l sec φ | φ | l sec φ |
|---|---|---|---|---|---|---|---|---|---|
| | **6.46** | | | **6.46** | | | **0.00** | | **0.00** |
| 0°00'.000 | | −∞ | 0°00'.000 | | −∞ | 0°00'.000 | 00 | 4°20'.701 | 00 |
| | 37 | | 0°44'.155 | 37 | 8.1087 | 0°52'.164 | | 4°30'.918 | 13 |
| 1°51'.479 | | 8.5108 | 1°40'.555 | 38 | 8.4663 | 1°30'.348 | 01 | 4°40'.762 | 14 |
| | 36 | | 2°15'.168 | 39 | 8.5948 | 1°56'.634 | 02 | 4°50'.271 | 15 |
| 2°49'.567 | | 8.6929 | 2°42'.563 | 40 | 8.6751 | 2°17'.998 | 03 | 4°59'.477 | 16 |
| | 35 | | 3°05'.959 | 41 | 8.7336 | 2°36'.469 | 04 | 5°08'.407 | 17 |
| 3°32'.313 | | 8.7904 | 3°26'.717 | 42 | 8.7796 | 2°52'.976 | 05 | 5°17'.084 | 18 |
| | 34 | | 3°45'.567 | 43 | 8.8176 | 3°08'.038 | 06 | 5°25'.528 | 19 |
| 4°07'.789 | | 8.8574 | 4°02'.954 | 44 | 8.8500 | 3°21'.977 | 07 | 5°33'.758 | 20 |
| | 33 | | 4°19'.171 | 45 | 8.8781 | 3°35'.016 | 08 | 5°41'.789 | 21 |
| 4°38'.783 | | 8.9085 | 4°34'.427 | 46 | 8.9031 | 3°47'.300 | 09 | 5°49'.633 | 22 |
| | 32 | | 4°48'.875 | 47 | 8.9255 | 3°58'.955 | 10 | 5°57'.305 | 23 |
| 5°06'.659 | | 8.9498 | 5°02'.628 | 48 | 8.9458 | 4°10'.064 | 11 | 6°04'.814 | 24 |
| | 31 | | 5°15'.780 | 49 | 8.9643 | 4°20'.701 | 12 | 6°12'.170 | 25 |
| 5°32'.201 | | 8.9845 | 5°28'.401 | 50 | 8.9815 | | | | |
| | 30 | | 5°40'.550 | 51 | 8.9973 | | | | |
| 5°55'.913 | | 9.0143 | 5°52'.278 | 52 | 9.0121 | | | | |
| | 29 | | 6°03'.623 | 53 | 9.0260 | | | | |
| 6°18'.138 | | 9.0405 | | | | | | | |

| φ | l sin φ | | l csc φ | l tan φ | | l ctn φ | l sec φ | | l cos φ | θ |
|---|---|---|---|---|---|---|---|---|---|---|
| 6° 00' | 9.0192 | 120 | 0.9808 | 9.0216 | 122 | 0.9784 | 0.0024 | 1 | 9.9976 | 84° 00' |
| 6° 10' | 9.0311 | 117 | 0.9689 | 9.0336 | 118 | 0.9664 | 0.0025 | 1 | 9.9975 | 83° 50' |
| 6° 20' | 9.0426 | 114 | 0.9574 | 9.0453 | 115 | 0.9547 | 0.0027 | 1 | 9.9973 | 83° 40' |
| 6° 30' | 9.0539 | 111 | 0.9461 | 9.0567 | 112 | 0.9433 | 0.0028 | 1 | 9.9972 | 83° 30' |
| 6° 40' | 9.0648 | 108 | 0.9352 | 9.0678 | 110 | 0.9322 | 0.0029 | 1 | 9.9971 | 83° 20' |
| 6° 50' | 9.0755 | 105 | 0.9245 | 9.0786 | 107 | 0.9214 | 0.0031 | 2 | 9.9969 | 83° 10' |
| 7° 00' | 9.0859 | 103 | 0.9141 | 9.0891 | 104 | 0.9109 | 0.0032 | 2 | 9.9968 | 83° 00' |
| 7° 10' | 9.0961 | 100 | 0.9039 | 9.0995 | 102 | 0.9005 | 0.0034 | 2 | 9.9966 | 82° 50' |
| 7° 20' | 9.1060 | 98 | 0.8940 | 9.1096 | 100 | 0.8904 | 0.0036 | 2 | 9.9964 | 82° 40' |
| 7° 30' | 9.1157 | 96 | 0.8843 | 9.1194 | 98 | 0.8806 | 0.0037 | 2 | 9.9963 | 82° 30' |
| 7° 40' | 9.1252 | 94 | 0.8748 | 9.1291 | 96 | 0.8709 | 0.0039 | 2 | 9.9961 | 82° 20' |
| 7° 50' | 9.1345 | 92 | 0.8655 | 9.1385 | 94 | 0.8615 | 0.0041 | 2 | 9.9959 | 82° 10' |
| 8° 00' | 9.1436 | 90 | 0.8564 | 9.1478 | 92 | 0.8522 | 0.0042 | 2 | 9.9958 | 82° 00' |
| 8° 10' | 9.1525 | 88 | 0.8475 | 9.1569 | 90 | 0.8431 | 0.0044 | 2 | 9.9956 | 81° 50' |
| 8° 20' | 9.1612 | 86 | 0.8388 | 9.1658 | 88 | 0.8342 | 0.0046 | 2 | 9.9954 | 81° 40' |
| 8° 30' | 9.1697 | 85 | 0.8303 | 9.1745 | 86 | 0.8255 | 0.0048 | 2 | 9.9952 | 81° 30' |
| 8° 40' | 9.1781 | 83 | 0.8219 | 9.1831 | 85 | 0.8169 | 0.0050 | 2 | 9.9950 | 81° 20' |
| 8° 50' | 9.1863 | 81 | 0.8137 | 9.1915 | 83 | 0.8085 | 0.0052 | 2 | 9.9948 | 81° 10' |
| 9° 00' | 9.1943 | 80 | 0.8057 | 9.1997 | 82 | 0.8003 | 0.0054 | 2 | 9.9946 | 81° 00' |
| 9° 10' | 9.2022 | 78 | 0.7978 | 9.2078 | 80 | 0.7922 | 0.0056 | 2 | 9.9944 | 80° 50' |
| 9° 20' | 9.2100 | 77 | 0.7900 | 9.2158 | 79 | 0.7842 | 0.0058 | 2 | 9.9942 | 80° 40' |
| 9° 30' | 9.2176 | 75 | 0.7824 | 9.2236 | 78 | 0.7764 | 0.0060 | 2 | 9.9940 | 80° 30' |
| 9° 40' | 9.2251 | 74 | 0.7749 | 9.2313 | 76 | 0.7687 | 0.0062 | 2 | 9.9938 | 80° 20' |
| 9° 50' | 9.2324 | 73 | 0.7676 | 9.2389 | 75 | 0.7611 | 0.0064 | 2 | 9.9936 | 80° 10' |
| 10° 00' | 9.2397 | 72 | 0.7603 | 9.2463 | 74 | 0.7537 | 0.0066 | 2 | 9.9934 | 80° 00' |
| | l cos θ | | l sec θ | l ctn θ | | l tan θ | l csc θ | | l sin θ | θ |

# Logarithms of Circular Functions.

| φ | l sin φ | l csc φ | l tan φ | l ctn φ | l sec φ | l cos φ | |
|---|---|---|---|---|---|---|---|
| **10° 00′** | 9.2397 72 | 0.7603 | 9.2463 74 | 0.7537 | 0.0066 2 | 9.9934 | **80° 00′** |
| 10° 10′ | 9.2468 70 | 0.7532 | 9.2536 73 | 0.7464 | 0.0069 2 | 9.9931 | 79° 50′ |
| 10° 20′ | 9.2538 69 | 0.7462 | 9.2609 72 | 0.7391 | 0.0071 2 | 9.9929 | 79° 40′ |
| 10° 30′ | 9.2606 68 | 0.7394 | 9.2680 71 | 0.7320 | 0.0073 2 | 9.9927 | 79° 30′ |
| 10° 40′ | 9.2674 67 | 0.7326 | 9.2750 69 | 0.7250 | 0.0076 2 | 9.9924 | 79° 20′ |
| 10° 50′ | 9.2740 66 | 0.7260 | 9.2819 68 | 0.7181 | 0.0078 2 | 9.9922 | 79° 10′ |
| **11° 00′** | 9.2806 65 | 0.7194 | 9.2887 67 | 0.7113 | 0.0081 2 | 9.9919 | **79° 00′** |
| 11° 10′ | 9.2870 64 | 0.7130 | 9.2953 66 | 0.7047 | 0.0083 2 | 9.9917 | 78° 50′ |
| 11° 20′ | 9.2934 63 | 0.7066 | 9.3020 66 | 0.6980 | 0.0086 3 | 9.9914 | 78° 40′ |
| 11° 30′ | 9.2997 62 | 0.7003 | 9.3085 65 | 0.6915 | 0.0088 3 | 9.9912 | 78° 30′ |
| 11° 40′ | 9.3058 61 | 0.6942 | 9.3149 64 | 0.6851 | 0.0091 3 | 9.9909 | 78° 20′ |
| 11° 50′ | 9.3119 60 | 0.6881 | 9.3212 63 | 0.6788 | 0.0093 3 | 9.9907 | 78° 10′ |
| **12° 00′** | 9.3179 59 | 0.6821 | 9.3275 62 | 0.6725 | 0.0096 3 | 9.9904 | **78° 00′** |
| 12° 10′ | 9.3238 59 | 0.6762 | 9.3336 61 | 0.6664 | 0.0099 3 | 9.9901 | 77° 50′ |
| 12° 20′ | 9.3296 58 | 0.6704 | 9.3397 61 | 0.6603 | 0.0101 3 | 9.9899 | 77° 40′ |
| 12° 30′ | 9.3353 57 | 0.6647 | 9.3458 60 | 0.6542 | 0.0104 3 | 9.9896 | 77° 30′ |
| 12° 40′ | 9.3410 56 | 0.6590 | 9.3517 59 | 0.6483 | 0.0107 3 | 9.9893 | 77° 20′ |
| 12° 50′ | 9.3466 55 | 0.6534 | 9.3576 58 | 0.6424 | 0.0110 3 | 9.9890 | 77° 10′ |
| **13° 00′** | 9.3521 55 | 0.6479 | 9.3634 58 | 0.6366 | 0.0113 3 | 9.9887 | **77° 00′** |
| 13° 10′ | 9.3575 54 | 0.6425 | 9.3691 57 | 0.6309 | 0.0116 3 | 9.9884 | 76° 50′ |
| 13° 20′ | 9.3629 53 | 0.6371 | 9.3748 56 | 0.6252 | 0.0119 3 | 9.9881 | 76° 40′ |
| 13° 30′ | 9.3682 53 | 0.6318 | 9.3804 56 | 0.6196 | 0.0122 3 | 9.9878 | 76° 30′ |
| 13° 40′ | 9.3734 52 | 0.6266 | 9.3859 55 | 0.6141 | 0.0125 3 | 9.9875 | 76° 20′ |
| 13° 50′ | 9.3786 51 | 0.6214 | 9.3914 54 | 0.6086 | 0.0128 3 | 9.9872 | 76° 10′ |
| **14° 00′** | 9.3837 51 | 0.6163 | 9.3968 54 | 0.6032 | 0.0131 3 | 9.9869 | **76° 00′** |
| 14° 10′ | 9.3887 50 | 0.6113 | 9.4021 53 | 0.5979 | 0.0134 3 | 9.9866 | 75° 50′ |
| 14° 20′ | 9.3937 49 | 0.6063 | 9.4074 53 | 0.5926 | 0.0137 3 | 9.9863 | 75° 40′ |
| 14° 30′ | 9.3986 49 | 0.6014 | 9.4127 52 | 0.5873 | 0.0141 3 | 9.9859 | 75° 30′ |
| 14° 40′ | 9.4035 48 | 0.5965 | 9.4178 52 | 0.5822 | 0.0144 3 | 9.9856 | 75° 20′ |
| 14° 50′ | 9.4083 48 | 0.5917 | 9.4230 51 | 0.5770 | 0.0147 3 | 9.9853 | 75° 10′ |
| **15° 00′** | 9.4130 47 | 0.5870 | 9.4281 51 | 0.5719 | 0.0151 3 | 9.9849 | **75° 00′** |
| 15° 10′ | 9.4177 47 | 0.5823 | 9.4331 50 | 0.5669 | 0.0154 3 | 9.9846 | 74° 50′ |
| 15° 20′ | 9.4223 46 | 0.5777 | 9.4381 50 | 0.5619 | 0.0157 3 | 9.9843 | 74° 40′ |
| 15° 30′ | 9.4269 46 | 0.5731 | 9.4430 49 | 0.5570 | 0.0161 4 | 9.9839 | 74° 30′ |
| 15° 40′ | 9.4314 45 | 0.5686 | 9.4479 49 | 0.5521 | 0.0164 4 | 9.9836 | 74° 20′ |
| 15° 50′ | 9.4359 45 | 0.5641 | 9.4527 48 | 0.5473 | 0.0168 4 | 9.9832 | 74° 10′ |
| **16° 00′** | 9.4403 44 | 0.5597 | 9.4575 48 | 0.5425 | 0.0172 4 | 9.9828 | **74° 00′** |
| 16° 10′ | 9.4447 44 | 0.5553 | 9.4622 47 | 0.5378 | 0.0175 4 | 9.9825 | 73° 50′ |
| 16° 20′ | 9.4491 43 | 0.5509 | 9.4669 47 | 0.5331 | 0.0179 4 | 9.9821 | 73° 40′ |
| 16° 30′ | 9.4533 43 | 0.5467 | 9.4716 46 | 0.5284 | 0.0183 4 | 9.9817 | 73° 30′ |
| 16° 40′ | 9.4576 42 | 0.5424 | 9.4762 46 | 0.5238 | 0.0186 4 | 9.9814 | 73° 20′ |
| 16° 50′ | 9.4618 42 | 0.5382 | 9.4808 46 | 0.5192 | 0.0190 4 | 9.9810 | 73° 10′ |
| **17° 00′** | 9.4659 41 | 0.5341 | 9.4853 45 | 0.5147 | 0.0194 4 | 9.9806 | **73° 00′** |
| | l cos θ | l sec θ | l ctn θ | l tan θ | l csc θ | l sin θ | θ |

# Logarithms of Circular Functions.

| φ | l sin φ | l csc φ | l tan φ | l ctn φ | l sec φ | l cos φ | |
|---|---|---|---|---|---|---|---|
| **17° 00'** | 9.4659 41 | 0.5341 | 9.4853 45 | 0.5147 | 0.0194 4 | 9.9806 | **73° 00'** |
| 17° 10' | 9.4700 41 | 0.5300 | 9.4898 45 | 0.5102 | 0.0198 4 | 9.9802 | 72° 50' |
| 17° 20' | 9.4741 40 | 0.5259 | 9.4943 44 | 0.5057 | 0.0202 4 | 9.9798 | 72° 40' |
| 17° 30' | 9.4781 40 | 0.5219 | 9.4987 44 | 0.5013 | 0.0206 4 | 9.9794 | 72° 30' |
| 17° 40' | 9.4821 40 | 0.5179 | 9.5031 44 | 0.4969 | 0.0210 4 | 9.9790 | 72° 20' |
| 17° 50' | 9.4861 39 | 0.5139 | 9.5075 43 | 0.4925 | 0.0214 4 | 9.9786 | 72° 10' |
| **18° 00'** | 9.4900 39 | 0.5100 | 9.5118 43 | 0.4882 | 0.0218 4 | 9.9782 | **72° 00'** |
| 18° 10' | 9.4939 38 | 0.5061 | 9.5161 43 | 0.4839 | 0.0222 4 | 9.9778 | 71° 50' |
| 18° 20' | 9.4977 38 | 0.5023 | 9.5203 42 | 0.4797 | 0.0226 4 | 9.9774 | 71° 40' |
| 18° 30' | 9.5015 38 | 0.4985 | 9.5245 42 | 0.4755 | 0.0230 4 | 9.9770 | 71° 30' |
| 18° 40' | 9.5052 37 | 0.4948 | 9.5287 42 | 0.4713 | 0.0235 4 | 9.9765 | 71° 20' |
| 18° 50' | 9.5090 37 | 0 4910 | 9.5329 41 | 0.4671 | 0.0239 4 | 9.9761 | 71° 10' |
| **19° 00'** | 9.5126 37 | 0.4874 | 9.5370 41 | 0.4630 | 0.0243 4 | 9.9757 | **71° 00'** |
| 19° 10' | 9.5163 36 | 0.4837 | 9.5411 41 | 0.4589 | 0.0248 4 | 9.9752 | 70° 50' |
| 19° 20' | 9.5199 36 | 0.4801 | 9.5451 40 | 0.4549 | 0.0252 4 | 9.9748 | 70° 40' |
| 19° 30' | 9.5235 36 | 0.4765 | 9.5491 40 | 0.4509 | 0.0257 4 | 9.9743 | 70° 30' |
| 19° 40' | 9.5270 35 | 0.4730 | 9.5531 40 | 0.4469 | 0.0261 5 | 9.9739 | 70° 20' |
| 19° 50' | 9.5306 35 | 0.4694 | 9.5571 40 | 0.4429 | 0.0266 5 | 9.9734 | 70° 10' |
| **20° 00'** | 9.5341 35 | 0.4659 | 9.5611 39 | 0.4389 | 0.0270 5 | 9.9730 | **70° 00'** |
| 20° 10' | 9.5375 34 | 0.4625 | 9.5650 39 | 0.4350 | 0.0275 5 | 9.9725 | 69° 50' |
| 20° 20' | 9.5409 34 | 0.4591 | 9.5689 39 | 0.4311 | 0.0279 5 | 9.9721 | 69° 40' |
| 20° 30' | 9.5443 34 | 0.4557 | 9.5727 39 | 0.4273 | 0.0284 5 | 9.9716 | 69° 30' |
| 20° 40' | 9.5477 33 | 0.4523 | 9.5766 38 | 0.4234 | 0.0289 5 | 9.9711 | 69° 20' |
| 20° 50' | 9.5510 33 | 0.4490 | 9.5804 38 | 0.4196 | 0.0294 5 | 9.9706 | 69° 10' |
| **21° 00'** | 9.5543 33 | 0.4457 | 9.5842 38 | 0.4158 | 0.0298 5 | 9.9702 | **69° 00'** |
| 21° 10' | 9.5576 33 | 0.4424 | 9.5879 38 | 0.4121 | 0.0303 5 | 9.9697 | 68° 50' |
| 21° 20' | 9.5609 32 | 0.4391 | 9.5917 37 | 0.4083 | 0.0308 5 | 9.9692 | 68° 40' |
| 21° 30' | 9.5641 32 | 0.4359 | 9.5954 37 | 0.4046 | 0.0313 5 | 9.9687 | 68° 30' |
| 21° 40' | 9.5673 32 | 0.4327 | 9.5991 37 | 0.4009 | 0.0318 5 | 9.9682 | 68° 20' |
| 21° 50' | 9.5704 32 | 0.4296 | 9.6028 37 | 0.3972 | 0.0323 5 | 9.9677 | 68° 10' |
| **22° 00'** | 9.5736 31 | 0.4264 | 9.6064 36 | 0.3936 | 0.0328 5 | 9.9672 | **68° 00'** |
| 22° 10' | 9.5767 31 | 0.4233 | 9.6100 36 | 0.3900 | 0.0333 5 | 9.9667 | 67° 50' |
| 22° 20' | 9.5798 31 | 0.4202 | 9.6136 36 | 0.3864 | 0.0339 5 | 9.9661 | 67° 40' |
| 22° 30' | 9.5828 30 | 0.4172 | 9.6172 36 | 0.3828 | 0.0344 5 | 9.9656 | 67 30' |
| 22° 40' | 9.5859 30 | 0.4141 | 9.6208 36 | 0.3792 | 0.0349 5 | 9.9651 | 67° 20' |
| 22° 50' | 9.5889 30 | 0.4111 | 9.6243 35 | 0.3757 | 0.0354 5 | 9.9646 | 67° 10' |
| **23° 00'** | 9.5919 30 | 0.4081 | 9.6279 35 | 0.3721 | 0.0360 5 | 9.9640 | **67° 00'** |
| 23° 10' | 9.5948 30 | 0.4052 | 9.6314 35 | 0.3686 | 0.0365 5 | 9.9635 | 66° 50' |
| 23° 20' | 9.5978 29 | 0.4022 | 9.6348 35 | 0.3652 | 0.0371 5 | 9.9629 | 66° 40' |
| 23° 30' | 9.6007 29 | 0.3993 | 9.6383 35 | 0.3617 | 0.0376 5 | 9.9624 | 66° 30' |
| 23° 40' | 9.6036 29 | 0.3964 | 9.6417 34 | 0.3583 | 0.0382 6 | 9.9618 | 66° 20' |
| 23° 50' | 9.6065 29 | 0.3935 | 9.6452 34 | 0.3548 | 0.0387 6 | 9.9613 | 66° 10' |
| **24° 00'** | 9.6093 28 | 0.3907 | 9.6486 34 | 0.3514 | 0.0393 6 | 9.9607 | **66° 00'** |
| | l cos θ | l sec θ | l ctn θ | l tan θ | l csc θ | l sin θ | θ |

## Logarithms of Circular Functions.

| $\phi$ | l sin $\phi$ | | l csc $\phi$ | l tan $\phi$ | | l ctn $\phi$ | l sec $\phi$ | | l cos $\phi$ | |
|---|---|---|---|---|---|---|---|---|---|---|
| **24° 00′** | 9.6093 | 28 | 0.3907 | 9.6486 | 34 | 0.3514 | 0.0393 | 6 | 9.9607 | **66° 00′** |
| 24° 10′ | 9.6121 | 28 | 0.3879 | 9.6520 | 34 | 0.3480 | 0.0398 | 6 | 9.9602 | 65° 50′ |
| 24° 20′ | 9.6149 | 28 | 0.3851 | 9.6553 | 34 | 0.3447 | 0.0404 | 6 | 9.9596 | 65° 40′ |
| 24° 30′ | 9.6177 | 28 | 0.3823 | 9.6587 | 33 | 0.3413 | 0.0410 | 6 | 9.9590 | 65° 30′ |
| 24° 40′ | 9.6205 | 28 | 0.3795 | 9.6620 | 33 | 0.3380 | 0.0416 | 6 | 9.9584 | 65° 20′ |
| 24° 50′ | 9.6232 | 27 | 0.3768 | 9.6654 | 33 | 0.3346 | 0.0421 | 6 | 9.9579 | 65° 10′ |
| **25° 00′** | 9.6259 | 27 | 0.3741 | 9.6687 | 33 | 0.3313 | 0.0427 | 6 | 9.9573 | **65° 00′** |
| 25° 10′ | 9.6286 | 27 | 0.3714 | 9.6720 | 33 | 0.3280 | 0.0433 | 6 | 9.9567 | 64° 50′ |
| 25° 20′ | 9.6313 | 27 | 0.3687 | 9.6752 | 33 | 0.3248 | 0.0439 | 6 | 9.9561 | 64° 40′ |
| 25° 30′ | 9.6340 | 26 | 0.3660 | 9.6785 | 33 | 0.3215 | 0.0445 | 6 | 9.9555 | 64° 30′ |
| 25° 40′ | 9.6366 | 26 | 0.3634 | 9.6817 | 32 | 0.3183 | 0.0451 | 6 | 9.9549 | 64° 20′ |
| 25° 50′ | 9.6392 | 26 | 0.3608 | 9.6850 | 32 | 0.3150 | 0.0457 | 6 | 9.9543 | 64° 10′ |
| **26° 00′** | 9.6418 | 26 | 0.3582 | 9.6882 | 32 | 0.3118 | 0.0463 | 6 | 9.9537 | **64° 00′** |
| 26° 10′ | 9.6444 | 26 | 0.3556 | 9.6914 | 32 | 0.3086 | 0.0470 | 6 | 9.9530 | 63° 50′ |
| 26° 20′ | 9.6470 | 26 | 0.3530 | 9.6946 | 32 | 0.3054 | 0.0476 | 6 | 9.9524 | 63° 40′ |
| 26° 30′ | 9.6495 | 25 | 0.3505 | 9.6977 | 32 | 0.3023 | 0.0482 | 6 | 9.9518 | 63° 30′ |
| 26° 40′ | 9.6521 | 25 | 0.3479 | 9.7009 | 31 | 0.2991 | 0.0488 | 6 | 9.9512 | 63° 20′ |
| 26° 50′ | 9.6546 | 25 | 0.3454 | 9.7040 | 31 | 0.2960 | 0.0495 | 6 | 9.9505 | 63° 10′ |
| **27° 00′** | 9.6570 | 25 | 0.3430 | 9.7072 | 31 | 0.2928 | 0.0501 | 6 | 9.9499 | **63° 00′** |
| 27° 10′ | 9.6595 | 25 | 0.3405 | 9.7103 | 31 | 0.2897 | 0.0508 | 6 | 9.9492 | 62° 50′ |
| 27° 20′ | 9.6620 | 24 | 0.3380 | 9.7134 | 31 | 0.2866 | 0.0514 | 7 | 9.9486 | 62° 40′ |
| 27° 30′ | 9.6644 | 24 | 0.3356 | 9.7165 | 31 | 0.2835 | 0.0521 | 7 | 9.9479 | 62° 30′ |
| 27° 40′ | 9.6668 | 24 | 0.3332 | 9.7196 | 31 | 0.2804 | 0.0527 | 7 | 9.9473 | 62° 20′ |
| 27° 50′ | 9.6692 | 24 | 0.3308 | 9.7226 | 31 | 0.2774 | 0.0534 | 7 | 9.9466 | 62° 10′ |
| **28° 00′** | 9.6716 | 24 | 0.3284 | 9.7257 | 30 | 0.2743 | 0.0541 | 7 | 9.9459 | **62° 00′** |
| 28° 10′ | 9.6740 | 24 | 0.3260 | 9.7287 | 30 | 0.2713 | 0.0547 | 7 | 9.9453 | 61° 50′ |
| 28° 20′ | 9.6763 | 23 | 0.3237 | 9.7317 | 30 | 0.2683 | 0.0554 | 7 | 9.9446 | 61° 40′ |
| 28° 30′ | 9.6787 | 23 | 0.3213 | 9.7348 | 30 | 0.2652 | 0.0561 | 7 | 9.9439 | 61° 30′ |
| 28° 40′ | 9.6810 | 23 | 0.3190 | 9.7378 | 30 | 0.2622 | 0.0568 | 7 | 9.9432 | 61° 20′ |
| 28° 50′ | 9.6833 | 23 | 0.3167 | 9.7408 | 30 | 0.2592 | 0.0575 | 7 | 9.9425 | 61° 10′ |
| **29° 00′** | 9.6856 | 23 | 0.3144 | 9.7438 | 30 | 0.2562 | 0.0582 | 7 | 9.9418 | **61° 00′** |
| 29° 10′ | 9.6878 | 23 | 0.3122 | 9.7467 | 30 | 0.2533 | 0.0589 | 7 | 9.9411 | 60° 50′ |
| 29° 20′ | 9.6901 | 22 | 0.3099 | 9.7497 | 30 | 0.2503 | 0.0596 | 7 | 9.9404 | 60° 40′ |
| 29° 30′ | 9.6923 | 22 | 0.3077 | 9.7526 | 29 | 0.2474 | 0.0603 | 7 | 9.9397 | 60° 30′ |
| 29° 40′ | 9.6946 | 22 | 0.3054 | 9.7556 | 29 | 0.2444 | 0.0610 | 7 | 9.9390 | 60° 20′ |
| 29° 50′ | 9.6968 | 22 | 0.3032 | 9.7585 | 29 | 0.2415 | 0.0617 | 7 | 9.9383 | 60° 10′ |
| **30° 00′** | 9.6990 | 22 | 0.3010 | 9.7614 | 29 | 0.2386 | 0.0625 | 7 | 9.9375 | **60° 00′** |
| 30° 10′ | 9.7012 | 22 | 0.2988 | 9.7644 | 29 | 0.2356 | 0.0632 | 7 | 9.9368 | 59° 50′ |
| 30° 20′ | 9.7033 | 22 | 0.2967 | 9.7673 | 29 | 0.2327 | 0.0639 | 7 | 9.9361 | 59° 40′ |
| 30° 30′ | 9.7055 | 21 | 0.2945 | 9.7701 | 29 | 0.2299 | 0.0647 | 7 | 9.9353 | 59° 30′ |
| 30° 40′ | 9.7076 | 21 | 0.2924 | 9.7730 | 29 | 0.2270 | 0.0654 | 7 | 9.9346 | 59° 20′ |
| 30° 50′ | 9.7097 | 21 | 0.2903 | 9.7759 | 29 | 0.2241 | 0.0662 | 8 | 9.9338 | 59° 10′ |
| **31° 00′** | 9.7118 | 21 | 0.2882 | 9.7788 | 29 | 0.2212 | 0.0669 | 8 | 9.9331 | **59° 00′** |
| | l cos $\theta$ | | l sec $\theta$ | l ctn $\theta$ | | l tan $\theta$ | l csc $\theta$ | | l sin $\theta$ | $\theta$ |

# Logarithms of Circular Functions.

| φ | lsin φ | | lcsc φ | ltan φ | | lctn φ | lsec φ | | lcos φ | |
|---|---|---|---|---|---|---|---|---|---|---|
| **31° 00′** | 9.7118 | 21 | 0.2882 | 9.7788 | 29 | 0.2212 | 0.0669 | 8 | 9.9331 | **59° 00′** |
| 31° 10′ | 9.7139 | 21 | 0.2861 | 9.7816 | 29 | 0.2184 | 0.0677 | 8 | 9.9323 | 58° 50′ |
| 31° 20′ | 9.7160 | 21 | 0.2840 | 9.7845 | 28 | 0.2155 | 0.0685 | 8 | 9.9315 | 58° 40′ |
| 31° 30′ | 9.7181 | 21 | 0.2819 | 9.7873 | 28 | 0.2127 | 0.0692 | 8 | 9.9308 | 58° 30′ |
| 31° 40′ | 9.7201 | 20 | 0.2799 | 9.7902 | 28 | 0.2098 | 0.0700 | 8 | 9.9300 | 58° 20′ |
| 31° 50′ | 9.7222 | 20 | 0.2778 | 9.7930 | 28 | 0.2070 | 0.0708 | 8 | 9.9292 | 58° 10′ |
| **32° 00′** | 9.7242 | 20 | 0.2758 | 9.7958 | 28 | 0.2042 | 0.0716 | 8 | 9.9284 | **58° 00′** |
| 32° 10′ | 9.7262 | 20 | 0.2738 | 9.7986 | 28 | 0.2014 | 0.0724 | 8 | 9.9276 | 57° 50′ |
| 32° 20′ | 9.7282 | 20 | 0.2718 | 9.8014 | 28 | 0.1986 | 0.0732 | 8 | 9.9268 | 57° 40′ |
| 32° 30′ | 9.7302 | 20 | 0.2698 | 9.8042 | 28 | 0.1958 | 0.0740 | 8 | 9.9260 | 57° 30′ |
| 32° 40′ | 9.7322 | 20 | 0.2678 | 9.8070 | 28 | 0.1930 | 0.0748 | 8 | 9.9252 | 57° 20′ |
| 32° 50′ | 9.7342 | 20 | 0.2658 | 9.8097 | 28 | 0.1903 | 0.0756 | 8 | 9.9244 | 57° 10′ |
| **33° 00′** | 9.7361 | 19 | 0.2639 | 9.8125 | 28 | 0.1875 | 0.0764 | 8 | 9.9236 | **57° 00′** |
| 33° 10′ | 9.7380 | 19 | 0.2620 | 9.8153 | 28 | 0.1847 | 0.0772 | 8 | 9.9228 | 56° 50′ |
| 33° 20′ | 9.7400 | 19 | 0.2600 | 9.8180 | 28 | 0.1820 | 0.0781 | 8 | 9.9219 | 56° 40′ |
| 33° 30′ | 9.7419 | 19 | 0.2581 | 9.8208 | 27 | 0.1792 | 0.0789 | 8 | 9.9211 | 56° 30′ |
| 33° 40′ | 9.7438 | 19 | 0.2562 | 9.8235 | 27 | 0.1765 | 0.0797 | 8 | 9.9203 | 56° 20′ |
| 33° 50′ | 9.7457 | 19 | 0.2543 | 9.8263 | 27 | 0.1737 | 0.0806 | 8 | 9.9194 | 56° 10′ |
| **34° 00′** | 9.7476 | 19 | 0.2524 | 9.8290 | 27 | 0.1710 | 0.0814 | 9 | 9.9186 | **56° 00′** |
| 34° 10′ | 9.7494 | 19 | 0.2506 | 9.8317 | 27 | 0.1683 | 0.0823 | 9 | 9.9177 | 55° 50′ |
| 34° 20′ | 9.7513 | 18 | 0.2487 | 9.8344 | 27 | 0.1656 | 0.0831 | 9 | 9.9169 | 55° 40′ |
| 34° 30′ | 9.7531 | 18 | 0.2469 | 9.8371 | 27 | 0.1629 | 0.0840 | 9 | 9.9160 | 55° 30′ |
| 34° 40′ | 9.7550 | 18 | 0.2450 | 9.8398 | 27 | 0.1602 | 0.0849 | 9 | 9.9151 | 55° 20′ |
| 34° 50′ | 9.7568 | 18 | 0.2432 | 9.8425 | 27 | 0.1575 | 0.0858 | 9 | 9.9142 | 55° 10′ |
| **35° 00′** | 9.7586 | 18 | 0.2414 | 9.8452 | 27 | 0.1548 | 0.0866 | 9 | 9.9134 | **55° 00′** |
| 35° 10′ | 9.7604 | 18 | 0.2396 | 9.8479 | 27 | 0.1521 | 0.0875 | 9 | 9.9125 | 54° 50′ |
| 35° 20′ | 9.7622 | 18 | 0.2378 | 9.8506 | 27 | 0.1494 | 0.0884 | 9 | 9.9116 | 54° 40′ |
| 35° 30′ | 9.7640 | 18 | 0.2360 | 9.8533 | 27 | 0.1467 | 0.0893 | 9 | 9.9107 | 54° 30′ |
| 35° 40′ | 9.7657 | 18 | 0.2343 | 9.8559 | 27 | 0.1441 | 0.0902 | 9 | 9.9098 | 54° 20′ |
| 35° 50′ | 9.7675 | 17 | 0.2325 | 9.8586 | 27 | 0.1414 | 0.0911 | 9 | 9.9089 | 54° 10′ |
| **36° 00′** | 9.7692 | 17 | 0.2308 | 9.8613 | 27 | 0.1387 | 0.0920 | 9 | 9.9080 | **54° 00′** |
| 36° 10′ | 9.7710 | 17 | 0.2290 | 9.8639 | 27 | 0.1361 | 0.0930 | 9 | 9.9070 | 53° 50′ |
| 36° 20′ | 9.7727 | 17 | 0.2273 | 9.8666 | 26 | 0.1334 | 0.0939 | 9 | 9.9061 | 53° 40′ |
| 36° 30′ | 9.7744 | 17 | 0.2256 | 9.8692 | 26 | 0.1308 | 0.0948 | 9 | 9.9052 | 53° 30′ |
| 36° 40′ | 9.7761 | 17 | 0.2239 | 9.8718 | 26 | 0.1282 | 0.0958 | 9 | 9.9042 | 53° 20′ |
| 36° 50′ | 9.7778 | 17 | 0.2222 | 9.8745 | 26 | 0.1255 | 0.0967 | 9 | 9.9033 | 53° 10′ |
| **37° 00′** | 9.7795 | 17 | 0.2205 | 9.8771 | 26 | 0.1229 | 0.0977 | 10 | 9.9023 | **53° 00′** |
| 37° 10′ | 9.7811 | 17 | 0.2189 | 9.8797 | 26 | 0.1203 | 0.0986 | 10 | 9.9014 | 52° 50′ |
| 37° 20′ | 9.7828 | 17 | 0.2172 | 9.8824 | 26 | 0.1176 | 0.0996 | 10 | 9.9004 | 52° 40′ |
| 37° 30′ | 9.7844 | 16 | 0.2156 | 9.8850 | 26 | 0.1150 | 0.1005 | 10 | 9.8995 | 52° 30′ |
| 37° 40′ | 9.7861 | 16 | 0.2139 | 9.8876 | 26 | 0.1124 | 0.1015 | 10 | 9.8985 | 52° 20′ |
| 37° 50′ | 9.7877 | 16 | 0.2123 | 9.8902 | 26 | 0.1098 | 0.1025 | 10 | 9.8975 | 52° 10′ |
| **38° 00′** | 9.7893 | 16 | 0.2107 | 9.8928 | 26 | 0.1072 | 0.1035 | 10 | 9.8965 | **52° 00′** |
| | lcos θ | | lsec θ | lctn θ | | ltan θ | lcsc θ | | lsin θ | θ |

| φ | l sin φ | l csc φ | l tan φ | l ctn φ | l sec φ | l cos φ | |
|---|---|---|---|---|---|---|---|
| **38° 00′** | 9.7893 16 | 0.2107 | 9.8928 26 | 0.1072 | 0.1035 10 | 9.8965 | **52° 00′** |
| 38° 10′ | 9.7910 16 | 0.2090 | 9.8954 26 | 0.1046 | 0.1045 10 | 9.8955 | 51° 50′ |
| 38° 20′ | 9.7926 16 | 0.2074 | 9.8980 26 | 0.1020 | 0.1055 10 | 9.8945 | 51° 40′ |
| 38° 30′ | 9.7941 16 | 0.2059 | 9.9006 26 | 0.0994 | 0.1065 10 | 9.8935 | 51° 30′ |
| 38° 40′ | 9.7957 16 | 0.2043 | 9.9032 26 | 0.0968 | 0.1075 10 | 9.8925 | 51° 20′ |
| 38° 50′ | 9.7973 16 | 0.2027 | 9.9058 26 | 0.0942 | 0.1085 10 | 9.8915 | 51° 10′ |
| **39° 00′** | 9.7989 16 | 0.2011 | 9.9084 26 | 0.0916 | 0.1095 10 | 9.8905 | **51° 00′** |
| 39° 10′ | 9.8004 16 | 0.1996 | 9.9110 26 | 0.0890 | 0.1105 10 | 9.8895 | 50° 50′ |
| 39° 20′ | 9.8020 15 | 0.1980 | 9.9135 26 | 0.0865 | 0.1116 10 | 9.8884 | 50° 40′ |
| 39° 30′ | 9.8035 15 | 0.1965 | 9.9161 26 | 0.0839 | 0.1126 10 | 9.8874 | 50° 30′ |
| 39° 40′ | 9.8050 15 | 0.1950 | 9.9187 26 | 0.0813 | 0.1136 10 | 9.8864 | 50° 20′ |
| 39° 50′ | 9.8066 15 | 0.1934 | 9.9212 26 | 0.0788 | 0.1147 11 | 9.8853 | 50° 10′ |
| **40° 00′** | 9.8081 15 | 0.1919 | 9.9238 26 | 0.0762 | 0.1157 11 | 9.8843 | **50° 00′** |
| 40° 10′ | 9.8096 15 | 0.1904 | 9.9264 26 | 0.0736 | 0.1168 11 | 9.8832 | 49° 50′ |
| 40° 20′ | 9.8111 15 | 0.1889 | 9.9289 26 | 0.0711 | 0.1179 11 | 9.8821 | 49° 40′ |
| 40° 30′ | 9.8125 15 | 0.1875 | 9.9315 26 | 0.0685 | 0.1190 11 | 9.8810 | 49° 30′ |
| 40° 40′ | 9.8140 15 | 0.1860 | 9.9341 26 | 0.0659 | 0.1200 11 | 9.8800 | 49° 20′ |
| 40° 50′ | 9.8155 15 | 0.1845 | 9.9366 26 | 0.0634 | 0.1211 11 | 9.8789 | 49° 10′ |
| **41° 00′** | 9.8169 15 | 0.1831 | 9.9392 26 | 0.0608 | 0.1222 11 | 9.8778 | **49° 00′** |
| 41° 10′ | 9.8184 14 | 0.1816 | 9.9417 25 | 0.0583 | 0.1233 11 | 9.8767 | 48° 50′ |
| 41° 20′ | 9.8198 14 | 0.1802 | 9.9443 25 | 0.0557 | 0.1244 11 | 9.8756 | 48° 40′ |
| 41° 30′ | 9.8213 14 | 0.1787 | 9.9468 25 | 0.0532 | 0.1255 11 | 9.8745 | 48° 30′ |
| 41° 40′ | 9.8227 14 | 0.1773 | 9.9494 25 | 0.0506 | 0.1267 11 | 9.8733 | 48° 20′ |
| 41° 50′ | 9.8241 14 | 0.1759 | 9.9519 25 | 0.0481 | 0.1278 11 | 9.8722 | 48° 10′ |
| **42° 00′** | 9.8255 14 | 0.1745 | 9.9544 25 | 0.0456 | 0.1289 11 | 9.8711 | **48° 00′** |
| 42° 10′ | 9.8269 14 | 0.1731 | 9.9570 25 | 0.0430 | 0.1301 11 | 9.8699 | 47° 50′ |
| 42° 20′ | 9.8283 14 | 0.1717 | 9.9595 25 | 0.0405 | 0.1312 12 | 9.8688 | 47° 40′ |
| 42° 30′ | 9.8297 14 | 0.1703 | 9.9621 25 | 0.0379 | 0.1324 12 | 9.8676 | 47° 30′ |
| 42° 40′ | 9.8311 14 | 0.1689 | 9.9646 25 | 0.0354 | 0.1335 12 | 9.8665 | 47° 20′ |
| 42° 50′ | 9.8324 14 | 0.1676 | 9.9671 25 | 0.0329 | 0.1347 12 | 9.8653 | 47° 10′ |
| **43° 00′** | 9.8338 14 | 0.1662 | 9.9697 25 | 0.0303 | 0.1359 12 | 9.8641 | **47° 00′** |
| 43° 10′ | 9.8351 13 | 0.1649 | 9.9722 25 | 0.0278 | 0.1371 12 | 9.8629 | 46° 50′ |
| 43° 20′ | 9.8365 13 | 0.1635 | 9.9747 25 | 0.0253 | 0.1382 12 | 9.8618 | 46° 40′ |
| 43° 30′ | 9.8378 13 | 0.1622 | 9.9772 25 | 0.0228 | 0.1394 12 | 9.8606 | 46° 30′ |
| 43° 40′ | 9.8391 13 | 0.1609 | 9.9798 25 | 0.0202 | 0.1406 12 | 9.8594 | 46° 20′ |
| 43° 50′ | 9.8405 13 | 0.1595 | 9.9823 25 | 0.0177 | 0.1418 12 | 9.8582 | 46° 10′ |
| **44° 00′** | 9.8418 13 | 0.1582 | 9.9848 25 | 0.0152 | 0.1431 12 | 9.8569 | **46° 00′** |
| 44° 10′ | 9.8431 13 | 0.1569 | 9.9874 25 | 0.0126 | 0.1443 12 | 9.8557 | 45° 50′ |
| 44° 20′ | 9.8444 13 | 0.1556 | 9.9899 25 | 0.0101 | 0.1455 12 | 9.8545 | 45° 40′ |
| 44° 30′ | 9.8457 13 | 0.1543 | 9.9924 25 | 0.0076 | 0.1468 12 | 9.8532 | 45° 30′ |
| 44° 40′ | 9.8469 13 | 0.1531 | 9.9949 25 | 0.0051 | 0.1480 12 | 9.8520 | 45° 20′ |
| 44° 50′ | 9.8482 13 | 0.1518 | 9.9975 25 | 0.0025 | 0.1493 13 | 9.8507 | 45° 10′ |
| **45° 00′** | 9.8495 13 | 0.1505 | 0.0000 25 | 0.0000 | 0.1505 13 | 9.8495 | **45° 00′** |
| | l cos θ | l sec θ | l ctn θ | l tan θ | l csc θ | l sin θ | θ |

| log $u$ | sin⁻¹$u$ cos⁻¹$u$ | tan⁻¹$u$ ctn⁻¹$u$ | log $u$ | sin⁻¹$u$ cos⁻¹$u$ | tan⁻¹$u$ ctn⁻¹$u$ |
|---|---|---|---|---|---|
| **9.** | ° ° | ° ° | **9.** | ° ° | ° ° |
| **00** | 5.74 13 84.26 | 5.71 13 84.29 | **50** | 18.43 44 71.57 | 17.55 38 72.45 |
| 01 | 5.87 14 84.13 | 5.84 13 84.16 | 51 | 18.88 45 71.12 | 17.93 39 72.07 |
| 02 | 6.01 14 83.99 | 5.98 14 84.02 | 52 | 19.34 46 70.66 | 18.32 39 71.68 |
| 03 | 6.15 14 83.85 | 6.12 14 83.88 | 53 | 19.81 48 70.19 | 18.72 40 71.28 |
| 04 | 6.30 15 83.70 | 6.26 14 83.74 | 54 | 20.29 49 69.71 | 19.12 41 70.88 |
| **05** | 6.44 15 83.56 | 6.40 15 83.60 | **55** | 20.78 50 69.22 | 19.54 42 70.46 |
| 06 | 6.59 15 83.41 | 6.55 15 83.45 | 56 | 21.29 51 68.71 | 19.95 42 70.05 |
| 07 | 6.75 16 83.25 | 6.70 15 83.30 | 57 | 21.81 53 68.19 | 20.38 43 69.62 |
| 08 | 6.91 16 83.09 | 6.86 16 83.14 | 58 | 22.35 54 67.65 | 20.82 44 69.18 |
| 09 | 7.07 16 82.93 | 7.01 16 82.99 | 59 | 22.90 56 67.10 | 21.26 45 68.74 |
| **10** | 7.23 17 82.77 | 7.18 16 82.82 | **60** | 23.46 57 66.54 | 21.71 45 68.29 |
| 11 | 7.40 17 82.60 | 7.34 17 82.66 | 61 | 24.04 59 65.96 | 22.17 46 67.83 |
| 12 | 7.58 18 82.42 | 7.51 17 82.49 | 62 | 24.64 61 65.36 | 22.63 47 67.37 |
| 13 | 7.75 18 82.25 | 7.68 17 82.32 | 63 | 25.25 62 64.75 | 23.10 48 66.90 |
| 14 | 7.93 18 82.07 | 7.86 18 82.14 | 64 | 25.88 64 64.12 | 23.58 48 66.42 |
| **15** | 8.12 19 81.88 | 8.04 18 81.96 | **65** | 26.53 66 63.47 | 24.07 49 65.93 |
| 16 | 8.31 19 81.69 | 8.22 19 81.78 | 66 | 27.20 68 62.80 | 24.56 50 65.44 |
| 17 | 8.51 20 81.49 | 8.41 19 81.59 | 67 | 27.89 70 62.11 | 25.07 51 64.93 |
| 18 | 8.71 20 81.29 | 8.61 20 81.39 | 68 | 28.60 72 61.40 | 25.58 51 64.42 |
| 19 | 8.91 21 81.09 | 8.80 20 81.20 | 69 | 29.33 74 60.67 | 26.09 52 63.91 |
| **20** | 9.12 21 80.88 | 9.01 20 80.99 | **70** | 30.08 76 59.92 | 26.62 53 63.38 |
| 21 | 9.33 22 80.67 | 9.21 21 80.79 | 71 | 30.85 79 59.15 | 27.15 54 62.85 |
| 22 | 9.55 22 80.45 | 9.42 21 80.58 | 72 | 31.66 81 58.34 | 27.69 54 62.31 |
| 23 | 9.78 23 80.22 | 9.64 22 80.36 | 73 | 32.48 84 57.52 | 28.24 55 61.76 |
| 24 | 10.01 23 79.99 | 9.86 22 80.14 | 74 | 33.34 87 56.66 | 28.79 56 61.21 |
| **25** | 10.24 24 79.76 | 10.08 23 79.92 | **75** | 34.22 90 55.78 | 29.35 56 60.65 |
| 26 | 10.48 24 79.52 | 10.31 23 79.69 | 76 | 35.13 54.87 | 29.92 57 60.08 |
| 27 | 10.73 25 79.27 | 10.55 24 79.45 | 77 | 36.07 · 53.93 | 30.49 58 59.51 |
| 28 | 10.98 26 79.02 | 10.79 24 79.21 | 78 | 37.05 52.95 | 31.07 58 58.93 |
| 29 | 11.24 26 78.76 | 11.03 25 78.97 | 79 | 38.07 51.93 | 31.66 59 58.34 |
| **30** | 11.51 27 78.49 | 11.28 25 78.72 | **80** | 39.12 50.88 | 32.25 60 57.75 |
| 31 | 11.78 28 78.22 | 11.54 26 78.46 | 81 | 40.21 49.79 | 32.85 60 57.15 |
| 32 | 12.06 28 77.94 | 11.80 26 78.20 | 82 | 41.35 48.65 | 33.45 61 56.55 |
| 33 | 12.34 29 77.66 | 12.07 27 77.93 | 83 | 42.54 47.46 | 34.06 61 55.94 |
| 34 | 12.64 30 77.36 | 12.34 28 77.66 | 84 | 43.78 46.22 | 34.68 62 55.32 |
| **35** | 12.94 30 77.06 | 12.62 28 77.38 | **85** | 45.07 44.93 | 35.30 62 54.70 |
| 36 | 13.24 31 76.76 | 12.90 29 77.10 | 86 | 46.42 43.58 | 35.92 63 54.08 |
| 37 | 13.56 32 76.44 | 13.19 29 76.81 | 87 | 47.84 42.16 | 36.55 63 53.45 |
| 38 | 13.88 33 76.12 | 13.49 30 76.51 | 88 | 49.31 40.66 | 37.18 64 52.82 |
| 39 | 14.21 33 75.79 | 13.79 31 76.21 | 89 | 50.92 39.08 | 37.82 64 52.18 |
| **40** | 14.55 34 75.45 | 14.10 31 75.90 | **90** | 52.59 37.41 | 38.46 64 51.54 |
| 41 | 14.89 35 75.11 | 14.42 32 75.58 | 91 | 54.37 35.63 | 39.11 65 50.89 |
| 42 | 15.25 36 74.75 | 14.74 32 75.26 | 92 | 56.28 33.72 | 39.75 65 50.25 |
| 43 | 15.61 37 74.39 | 15.06 33 74.94 | 93 | 58.34 31.66 | 40.40 65 49.60 |
| 44 | 15.99 38 74.01 | 15.40 34 74.60 | 94 | 60.57 29.43 | 41.05 65 48.95 |
| **45** | 16.37 39 73.63 | 15.74 34 74.26 | **95** | 63.03 26.97 | 41.71 66 48.29 |
| 46 | 16.76 40 73.24 | 16.09 35 73.91 | 96 | 65.78 24.22 | 42.37 66 47.63 |
| 47 | 17.16 41 72.84 | 16.44 36 73.56 | 97 | 68.95 21.05 | 43.02 66 46.98 |
| 48 | 17.58 42 72.42 | 16.80 37 73.20 | 98 | 72.74 17.26 | 43.68 66 46.32 |
| 49 | 18.00 43 72.00 | 17.17 37 72.83 | 99 | 77.75 12.25 | 44.34 66 45.66 |
| **50** | 18.43 44 71.57 | 17.55 38 72.45 | **00** | 90.00 00.00 | 45.00 66 45.00 |

# Inverse Circular Functions.

| log u | sin⁻¹ u | | cos⁻¹ u | log u | sin⁻¹ u | | cos⁻¹ u | log u | sin⁻¹ u | | cos⁻¹ u |
|---|---|---|---|---|---|---|---|---|---|---|---|
| **9.** | ° | | ° | **9.** | ° | | ° | **9.** | ° | | ° |
| **750** | 34.22 | 9 | 55.78 | **800** | 39.12 | 11 | 50.88 | **850** | 45.07 | 13 | 44.93 |
| 751 | 34.31 | 9 | 55.69 | 801 | 39.23 | 11 | 50.77 | 851 | 45.20 | 13 | 44.80 |
| 752 | 34.40 | 9 | 55.60 | 802 | 39.34 | 11 | 50.66 | 852 | 45.33 | 13 | 44.67 |
| 753 | 34.49 | 9 | 55.51 | 803 | 39.44 | 11 | 50.56 | 853 | 45.47 | 13 | 44.53 |
| 754 | 34.58 | 9 | 55.42 | 804 | 39.55 | 11 | 50.45 | 854 | 45.60 | 13 | 44.40 |
| **755** | 34.67 | 9 | 55.33 | **805** | 39.66 | 11 | 50.34 | **855** | 45.74 | 14 | 44.26 |
| 756 | 34.76 | 9 | 55.24 | 806 | 39.77 | 11 | 50.23 | 856 | 45.87 | 14 | 44.13 |
| 757 | 34.85 | 9 | 55.15 | 807 | 39.88 | 11 | 50.12 | 857 | 46.01 | 14 | 43.99 |
| 758 | 34.95 | 9 | 55.05 | 808 | 39.99 | 11 | 50.01 | 858 | 46.15 | 14 | 43.85 |
| 759 | 35.04 | 9 | 54.96 | 809 | 40.10 | 11 | 49.90 | 859 | 46.28 | 14 | 43.72 |
| **760** | 35.13 | 9 | 54.87 | **810** | 40.21 | 11 | 49.79 | **860** | 46.42 | 14 | 43.58 |
| 761 | 35.22 | 9 | 54.78 | 811 | 40.33 | 11 | 49.67 | 861 | 46.56 | 14 | 43.44 |
| 762 | 35.32 | 9 | 54.68 | 812 | 40.44 | 11 | 49.56 | 862 | 46.70 | 14 | 43.30 |
| 763 | 35.41 | 9 | 54.59 | 813 | 40.55 | 11 | 49.45 | 863 | 46.84 | 14 | 43.16 |
| 764 | 35.50 | 9 | 54.50 | 814 | 40.66 | 11 | 49.34 | 864 | 46.98 | 14 | 43.02 |
| **765** | 35.60 | 9 | 54.40 | **815** | 40.78 | 11 | 49.22 | **865** | 47.12 | 14 | 42.88 |
| 766 | 35.69 | 9 | 54.31 | 816 | 40.89 | 11 | 49.11 | 866 | 47.27 | 14 | 42.73 |
| 767 | 35.79 | 10 | 54.21 | 817 | 41.01 | 11 | 48.99 | 867 | 47.41 | 14 | 42.59 |
| 768 | 35.88 | 10 | 54.12 | 818 | 41.12 | 12 | 48.88 | 868 | 47.55 | 14 | 42.45 |
| 769 | 35.98 | 10 | 54.02 | 819 | 41.24 | 12 | 48.76 | 869 | 47.70 | 14 | 42.30 |
| **770** | 36.07 | 10 | 53.93 | **820** | 41.35 | 12 | 48.65 | **870** | 47.84 | 15 | 42.16 |
| 771 | 36.17 | 10 | 53.83 | 821 | 41.47 | 12 | 48.53 | 871 | 47.99 | 15 | 42.01 |
| 772 | 36.27 | 10 | 53.73 | 822 | 41.59 | 12 | 48.41 | 872 | 48.14 | 15 | 41.86 |
| 773 | 36.36 | 10 | 53.64 | 823 | 41.70 | 12 | 48.30 | 873 | 48.28 | 15 | 41.72 |
| 774 | 36.46 | 10 | 53.54 | 824 | 41.82 | 12 | 48.18 | 874 | 48.43 | 15 | 41.57 |
| **775** | 36.56 | 10 | 53.44 | **825** | 41.94 | 12 | 48.06 | **875** | 48.58 | 15 | 41.42 |
| 776 | 36.66 | 10 | 53.34 | 826 | 42.06 | 12 | 47.94 | 876 | 48.73 | 15 | 41.27 |
| 777 | 36.76 | 10 | 53.24 | 827 | 42.18 | 12 | 47.82 | 877 | 48.88 | 15 | 41.12 |
| 778 | 36.85 | 10 | 53.15 | 828 | 42.30 | 12 | 47.70 | 878 | 49.03 | 15 | 40.97 |
| 779 | 36.95 | 10 | 53.05 | 829 | 42.42 | 12 | 47.58 | 879 | 49.19 | 15 | 40.81 |
| **780** | 37.05 | 10 | 52.95 | **830** | 42.54 | 12 | 47.46 | **880** | 49.34 | 15 | 40.66 |
| 781 | 37.15 | 10 | 52.85 | 831 | 42.66 | 12 | 47.34 | 881 | 49.49 | 15 | 40.51 |
| 782 | 37.25 | 10 | 52.75 | 832 | 42.78 | 12 | 47.22 | 882 | 49.65 | 16 | 40.35 |
| 783 | 37.35 | 10 | 52.65 | 833 | 42.90 | 12 | 47.10 | 883 | 49.80 | 16 | 40.20 |
| 784 | 37.45 | 10 | 52.55 | 834 | 43.03 | 12 | 46.97 | 884 | 49.96 | 16 | 40.04 |
| **785** | 37.56 | 10 | 52.44 | **835** | 43.15 | 12 | 46.85 | **885** | 50.12 | 16 | 39.88 |
| 786 | 37.66 | 10 | 52.34 | 836 | 43.27 | 12 | 46.73 | 886 | 50.28 | 16 | 39.72 |
| 787 | 37.76 | 10 | 52.24 | 837 | 43.40 | 12 | 46.60 | 887 | 50.44 | 16 | 39.56 |
| 788 | 37.86 | 10 | 52.14 | 838 | 43.52 | 13 | 46.48 | 888 | 50.60 | 16 | 39.40 |
| 789 | 37.96 | 10 | 52.04 | 839 | 43.65 | 13 | 46.35 | 889 | 50.76 | 16 | 39.24 |
| **790** | 38.07 | 10 | 51.93 | **840** | 43.78 | 13 | 46.22 | **890** | 50.92 | 16 | 39.08 |
| 791 | 38.17 | 10 | 51.83 | 841 | 43.90 | 13 | 46.10 | 891 | 51.08 | 16 | 38.92 |
| 792 | 38.28 | 10 | 51.72 | 842 | 44.03 | 13 | 45.97 | 892 | 51.25 | 16 | 38.75 |
| 793 | 38.38 | 10 | 51.62 | 843 | 44.16 | 13 | 45.84 | 893 | 51.41 | 17 | 38.59 |
| 794 | 38.48 | 10 | 51.52 | 844 | 44.29 | 13 | 45.71 | 894 | 51.58 | 17 | 38.42 |
| **795** | 38.59 | 11 | 51.41 | **845** | 44.41 | 13 | 45.59 | **895** | 51.74 | 17 | 38.26 |
| 796 | 38.69 | 11 | 51.31 | 846 | 44.54 | 13 | 45.46 | 896 | 51.91 | 17 | 38.09 |
| 797 | 38.80 | 11 | 51.20 | 847 | 44.67 | 13 | 45.33 | 897 | 52.08 | 17 | 37.92 |
| 798 | 38.91 | 11 | 51.09 | 848 | 44.80 | 13 | 45.20 | 898 | 52.25 | 17 | 37.75 |
| 799 | 39.01 | 11 | 50.99 | 849 | 44.94 | 13 | 45.06 | 899 | 52.42 | 17 | 37.58 |
| **800** | 39.12 | 11 | 50.88 | **850** | 45.07 | 13 | 44.93 | **900** | 52.59 | 17 | 37.41 |

# Inverse Circular Functions.

| log u | sin⁻¹u | | cos⁻¹u | log u | sin⁻¹u | | cos⁻¹u | log u | sin⁻¹u | log u | sin⁻¹u |
|---|---|---|---|---|---|---|---|---|---|---|---|
| **9.** | ° | | ° | **9.** | ° | | ° | **9.** | ° | **9.** | ° |
| **900** | 52.59 | 17 | 37.41 | **950** | 63.03 | 26 | 26.97 | **9900** | 77.75 | **9950** | 81.32 |
| 901 | 52.76 | 17 | 37.24 | 951 | 63.29 | 26 | 26.71 | 9901 | 77.81 | 9951 | 81.41 |
| 902 | 52.94 | 17 | 37.06 | 952 | 63.56 | 27 | 26.44 | 9902 | 77.87 | 9952 | 81.50 |
| 903 | 53.11 | 18 | 36.89 | 953 | 63.82 | 27 | 26.18 | 9903 | 77.94 | 9953 | 81.59 |
| 904 | 53.29 | 18 | 36.71 | 954 | 64.09 | 27 | 25.91 | 9904 | 78.00 | 9954 | 81.68 |
| **905** | 53.47 | 18 | 36.53 | **955** | 64.37 | 27 | 25.63 | **9905** | 78.06 | **9955** | 81.77 |
| 906 | 53.65 | 18 | 36.35 | 956 | 64.64 | 28 | 25.36 | 9906 | 78.12 | 9956 | 81.86 |
| 907 | 53.83 | 18 | 36.17 | 957 | 64.92 | 28 | 25.08 | 9907 | 78.18 | 9957 | 81.95 |
| 908 | 54.01 | 18 | 35.99 | 958 | 65.21 | 29 | 24.79 | 9908 | 78.25 | 9958 | 82.04 |
| 909 | 54.19 | 18 | 35.81 | 959 | 65.49 | 29 | 24.51 | 9909 | 78.31 | 9959 | 82.14 |
| **910** | 54.37 | 18 | 35.63 | **960** | 65.78 | 29 | 24.22 | **9910** | 78.38 | **9960** | 82.24 |
| 911 | 54.56 | 19 | 35.44 | 961 | 66.08 | 30 | 23.92 | 9911 | 78.44 | 9961 | 82.33 |
| 912 | 54.74 | 19 | 35.26 | 962 | 66.38 | 30 | 23.62 | 9912 | 78.50 | 9962 | 82.43 |
| 913 | 54.93 | 19 | 35.07 | 963 | 66.68 | 31 | 23.32 | 9913 | 78.57 | 9963 | 82.5 |
| 914 | 55.12 | 19 | 34.88 | 964 | 66.99 | 31 | 23.01 | 9914 | 78.64 | 9964 | 82.6 |
| **915** | 55.31 | 19 | 34.69 | **965** | 67.30 | 32 | 22.70 | **9915** | 78.70 | **9965** | 82.7 |
| 916 | 55.50 | 19 | 34.50 | 966 | 67.62 | 32 | 22.38 | 9916 | 78.77 | 9966 | 82.8 |
| 917 | 55.69 | 19 | 34.31 | 967 | 67.95 | 33 | 22.05 | 9917 | 78.83 | 9967 | 82.9 |
| 918 | 55.89 | 19 | 34.11 | 968 | 68.27 | 33 | 21.73 | 9918 | 78.90 | 9968 | 83.1 |
| 919 | 56.08 | 20 | 33.92 | 969 | 68.61 | 34 | 21.39 | 9919 | 78.97 | 9969 | 83.2 |
| **920** | 56.28 | 20 | 33.72 | **970** | 68.95 | 34 | 21.05 | **9920** | 79.04 | **9970** | 83.3 |
| 921 | 56.48 | 20 | 33.52 | 971 | 69.29 | 35 | 20.71 | 9921 | 79.10 | 9971 | 83.4 |
| 922 | 56.68 | 20 | 33.32 | 972 | 69.65 | 36 | 20.35 | 9922 | 79.17 | 9972 | 83.5 |
| 923 | 56.88 | 20 | 33.12 | 973 | 70.01 | 36 | 19.99 | 9923 | 79.24 | 9973 | 83.6 |
| 924 | 57.08 | 20 | 32.92 | 974 | 70.37 | 37 | 19.63 | 9924 | 79.31 | 9974 | 83.7 |
| **925** | 57.29 | 21 | 32.71 | **975** | 70.75 | 38 | 19.25 | **9925** | 79.38 | **9975** | 83.9 |
| 926 | 57.49 | 21 | 32.51 | 976 | 71.13 | 39 | 18.87 | 9926 | 79.45 | 9976 | 84.0 |
| 927 | 57.70 | 21 | 32.30 | 977 | 71.52 | 39 | 18.48 | 9927 | 79.52 | 9977 | 84.1 |
| 928 | 57.91 | 21 | 32.09 | 978 | 71.92 | 40 | 18.08 | 9928 | 79.60 | 9978 | 84.2 |
| 929 | 58.12 | 21 | 31.88 | 979 | 72.33 | 41 | 17.67 | 9929 | 79.67 | 9979 | 84.4 |
| **930** | 58.34 | 21 | 31.66 | **980** | 72.74 | 42 | 17.26 | **9930** | 79.74 | **9980** | 84.5 |
| 931 | 58.55 | 22 | 31.45 | 981 | 73.18 | 44 | 16.82 | 9931 | 79.81 | 9981 | 84.6 |
| 932 | 58.77 | 22 | 31.23 | 982 | 73.62 | 45 | 16.38 | 9932 | 79.89 | 9982 | 84.8 |
| 933 | 58.99 | 22 | 31.01 | 983 | 74.07 | 46 | 15.93 | 9933 | 79.96 | 9983 | 84.9 |
| 934 | 59.21 | 22 | 30.79 | 984 | 74.54 | 48 | 15.46 | 9934 | 80.04 | 9984 | 85.1 |
| **935** | 59.43 | 22 | 30.57 | **985** | 75.03 | 49 | 14.97 | **9935** | 80.11 | **9985** | 85.2 |
| 936 | 59.65 | 23 | 30.35 | 986 | 75.53 | 51 | 14.47 | 9936 | 80.19 | 9986 | 85.4 |
| 937 | 59.88 | 23 | 30.12 | 987 | 76.05 | 53 | 13.95 | 9937 | 80.26 | 9987 | 85.6 |
| 938 | 60.11 | 23 | 29.89 | 988 | 76.59 | 55 | 13.41 | 9938 | 80.34 | 9988 | 85.7 |
| 939 | 60.34 | 23 | 29.66 | 989 | 77.16 | 58 | 12.84 | 9939 | 80.42 | 9989 | 85.9 |
| **940** | 60.57 | 23 | 29.43 | **990** | 77.75 | 61 | 12.25 | **9940** | 80.50 | **9990** | 86.1 |
| 941 | 60.81 | 24 | 29.19 | 991 | 78.38 | | 11.62 | 9941 | 80.58 | 9991 | 86.3 |
| 942 | 61.04 | 24 | 28.96 | 992 | 79.04 | | 10.96 | 9942 | 80.66 | 9992 | 86.5 |
| 943 | 61.28 | 24 | 28.72 | 993 | 79.74 | | 10.26 | 9943 | 80.74 | 9993 | 86.7 |
| 944 | 61.52 | 24 | 28.48 | 994 | 80.50 | | 9.50 | 9944 | 80.82 | 9994 | 87.0 |
| **945** | 61.77 | 25 | 28.23 | **995** | 81.32 | | 8.68 | **9945** | 80.90 | **9995** | 87.3 |
| 946 | 62.02 | 25 | 27.98 | 996 | 82.24 | | 7.76 | 9946 | 80.98 | 9996 | 87.5 |
| 947 | 62.27 | 25 | 27.73 | 997 | 83.3 | | 6.7 | 9947 | 81.07 | 9997 | 87.9 |
| 948 | 62.52 | 25 | 27.48 | 998 | 84.5 | | 5.5 | 9948 | 81.15 | 9998 | 88.3 |
| 949 | 62.77 | 26 | 27.23 | 999 | 86.1 | | 3.9 | 9949 | 81.24 | 9999 | 88.8 |
| **950** | 63.03 | 26 | 26.97 | **000** | 90. | | 0. | **9950** | 81.32 | **0000** | 90. |

# Logarithms of Hyperbolic Functions.

| x | gd x | x | gd x | x | l Sh x | l Ch x | l Th x |
|---|---|---|---|---|---|---|---|
| | ° | | ° | | 0. | C. | 9. |
| 0.00 | 0.0000 5730 | 0.50 | 27.524 508 | 1.00 | 0701 57 | 1864 33 | 8817 24 |
| 0.01 | 0.5729 5729 | 0.51 | 28.031 506 | 1.01 | 0758 57 | 1917 33 | 8840 23 |
| 0.02 | 1.1458 5728 | 0.52 | 28.535 503 | 1.02 | 0815 56 | 1950 33 | 8864 23 |
| 0.03 | 1.7186 5727 | 0.53 | 29.037 501 | 1.03 | 0871 56 | 1984 34 | 8887 23 |
| 0.04 | 2.2912 5725 | 0.54 | 29.537 498 | 1.04 | 0927 56 | 2018 34 | 8909 22 |
| 0.05 | 2.8636 5722 | 0.55 | 30.034 496 | 1.05 | 0982 56 | 2051 34 | 8931 22 |
| 0.06 | 3.4357 5719 | 0.56 | 30.529 494 | 1.06 | 1038 55 | 2086 34 | 8952 21 |
| 0.07 | 4.0074 5716 | 0.57 | 31.021 491 | 1.07 | 1093 55 | 2120 34 | 8973 21 |
| 0.08 | 4.5788 5711 | 0.58 | 31.511 488 | 1.08 | 1148 55 | 2154 34 | 8994 20 |
| 0.09 | 5.1497 5706 | 0.59 | 31.998 486 | 1.09 | 1203 54 | 2189 35 | 9014 20 |
| .10 | 5.720 570 | 0.60 | 32.483 483 | 1.10 | 1257 54 | 2223 35 | 9034 19 |
| 0.11 | 6.290 570 | 0.61 | 32.965 481 | 1.11 | 1311 54 | 2258 35 | 9053 19 |
| 0.12 | 6.859 569 | 0.62 | 33.444 478 | 1.12 | 1365 54 | 2293 35 | 9072 19 |
| 0.13 | 7.428 568 | 0.63 | 33.921 475 | 1.13 | 1419 54 | 2328 35 | 9090 18 |
| 0.14 | 7.995 567 | 0.64 | 34.395 473 | 1.14 | 1472 53 | 2364 35 | 9108 18 |
| 0.15 | 8.562 567 | 0.65 | 34.867 470 | 1.15 | 1525 53 | 2399 36 | 9126 18 |
| 0.16 | 9.128 566 | 0.66 | 35.336 467 | 1.16 | 1578 53 | 2435 36 | 9144 17 |
| 0.17 | 9.694 565 | 0.67 | 35.802 465 | 1.17 | 1631 53 | 2470 36 | 9161 17 |
| 0.18 | 10.258 564 | 0.68 | 36.265 462 | 1.18 | 1684 52 | 2506 36 | 9177 17 |
| 0.19 | 10.821 563 | 0.69 | 36.726 459 | 1.19 | 1736 52 | 2542 36 | 9194 16 |
| 0.20 | 11.384 562 | 0.70 | 37.183 456 | 1.20 | 1788 52 | 2578 36 | 9210 16 |
| 0.21 | 11.945 561 | 0.71 | 37.638 454 | 1.21 | 1840 52 | 2615 36 | 9226 16 |
| 0.22 | 12.505 559 | 0.72 | 38.091 451 | 1.22 | 1892 52 | 2651 36 | 9241 15 |
| 0.23 | 13.063 558 | 0.73 | 38.540 448 | 1.23 | 1944 52 | 2688 37 | 9256 15 |
| 0.24 | 13.621 557 | 0.74 | 38.987 445 | 1.24 | 1995 51 | 2724 37 | 9271 15 |
| 0.25 | 14.177 556 | 0.75 | 39.431 443 | 1.25 | 2046 51 | 2761 37 | 9285 14 |
| 0.26 | 14.732 554 | 0.76 | 39.872 440 | 1.26 | 2098 51 | 2798 37 | 9300 14 |
| 0.27 | 15.286 553 | 0.77 | 40.310 437 | 1.27 | 2148 51 | 2835 37 | 9314 14 |
| 0.28 | 15.837 551 | 0.78 | 40.746 434 | 1.28 | 2199 51 | 2872 37 | 9327 14 |
| 0.29 | 16.388 550 | 0.79 | 41.179 431 | 1.29 | 2250 51 | 2909 37 | 9341 13 |
| 0.30 | 16.937 548 | 0.80 | 41.608 428 | 1.30 | 2300 50 | 2947 37 | 9354 13 |
| 0.31 | 17.484 546 | 0.81 | 42.035 426 | 1.31 | 2351 50 | 2984 38 | 9367 13 |
| 0.32 | 18.030 545 | 0.82 | 42.460 423 | 1.32 | 2401 50 | 3022 38 | 9379 12 |
| 0.33 | 18.573 543 | 0.83 | 42.881 420 | 1.33 | 2451 50 | 3059 38 | 9391 12 |
| 0.34 | 19.116 541 | 0.84 | 43.299 417 | 1.34 | 2501 50 | 3097 38 | 9404 12 |
| 0.35 | 19.656 540 | 0.85 | 43.715 414 | 1.35 | 2551 50 | 3135 38 | 9415 12 |
| 0.36 | 20.195 538 | 0.86 | 44.128 411 | 1.36 | 2600 50 | 3173 38 | 9427 11 |
| 0.37 | 20.732 536 | 0.87 | 44.537 408 | 1.37 | 2650 49 | 3211 38 | 9438 11 |
| 0.38 | 21.267 534 | 0.88 | 44.944 406 | 1.38 | 2699 49 | 3249 38 | 9450 11 |
| 0.39 | 21.800 532 | 0.89 | 45.348 403 | 1.39 | 2748 49 | 3288 38 | 9460 11 |
| 0.40 | 22.331 530 | 0.90 | 45.750 400 | 1.40 | 2797 49 | 3326 38 | 9471 11 |
| 0.41 | 22.859 528 | 0.91 | 46.148 397 | 1.41 | 2846 49 | 3365 39 | 9482 10 |
| 0.42 | 23.386 526 | 0.92 | 46.544 394 | 1.42 | 2895 49 | 3403 39 | 9492 10 |
| 0.43 | 23.911 524 | 0.93 | 46.936 391 | 1.43 | 2944 49 | 3442 39 | 9502 10 |
| 0.44 | 24.434 522 | 0.94 | 47.326 388 | 1.44 | 2993 49 | 3481 39 | 9512 10 |
| 0.45 | 24.955 519 | 0.95 | 47.713 386 | 1.45 | 3041 48 | 3520 39 | 9522 10 |
| 0.46 | 25.473 517 | 0.96 | 48.097 383 | 1.46 | 3090 48 | 3559 39 | 9531 9 |
| 0.47 | 25.989 515 | 0.97 | 48.478 380 | 1.47 | 3138 48 | 3598 39 | 9540 9 |
| 0.48 | 26.503 513 | 0.98 | 48.857 377 | 1.48 | 3186 48 | 3637 39 | 9549 9 |
| 0.49 | 27.015 510 | 0.99 | 49.232 374 | 1.49 | 3234 48 | 3676 39 | 9558 9 |
| 0.50 | 27.524 508 | 1.00 | 49.605 371 | 1.50 | 3282 48 | 3715 39 | 9567 9 |

# Logarithms of Hyperbolic Functions.

| $x$ | l Sh $x$ | l Ch $x$ | l Th $x$ | $x$ | l Sh $x$ | l Ch $x$ | l Th $x$ |
|---|---|---|---|---|---|---|---|
| | 0. | 0. | 9. | | 0. | 0. | 9. |
| **1.50** | 3282 48 | 3715 39 | 9567 9 | **2.00** | 5595 45 | 5754 42 | 9841 3 |
| 1.51 | 3330 48 | 3754 39 | 9576 8 | 2.01 | 5640 45 | 5796 42 | 9844 3 |
| 1.52 | 3378 48 | 3794 39 | 9584 8 | 2.02 | 5685 45 | 5838 42 | 9847 3 |
| 1.53 | 3426 48 | 3833 40 | 9592 8 | 2.03 | 5730 45 | 5880 42 | 9850 3 |
| 1.54 | 3474 48 | 3873 40 | 9601 8 | 2.04 | 5775 45 | 5922 42 | 9853 3 |
| **1.55** | 3521 48 | 3913 40 | 9608 8 | **2.05** | 5820 45 | 5964 42 | 9856 3 |
| 1.56 | 3569 47 | 3952 40 | 9616 8 | 2.06 | 5865 45 | 6006 42 | 9859 3 |
| 1.57 | 3616 47 | 3992 40 | 9624 8 | 2.07 | 5910 45 | 6048 42 | 9862 3 |
| 1.58 | 3663 47 | 4032 40 | 9631 7 | 2.08 | 5955 45 | 6090 42 | 9864 3 |
| 1.59 | 3711 47 | 4072 40 | 9639 7 | 2.09 | 6000 45 | 6132 42 | 9867 3 |
| **1.60** | 3758 47 | 4112 40 | 9646 7 | **2.10** | 6044 45 | 6175 42 | 9870 3 |
| 1.61 | 3805 47 | 4152 40 | 9653 7 | 2.11 | 6089 45 | 6217 42 | 9872 3 |
| 1.62 | 3852 47 | 4192 40 | 9660 7 | 2.12 | 6134 45 | 6259 42 | 9875 3 |
| 1.63 | 3899 47 | 4232 40 | 9666 7 | 2.13 | 6178 45 | 6301 42 | 9877 2 |
| 1.64 | 3946 47 | 4273 40 | 9673 7 | 2.14 | 6223 45 | 6343 42 | 9880 2 |
| **1.65** | 3992 47 | 4313 40 | 9679 6 | **2.15** | 6268 45 | 6386 42 | 9882 2 |
| 1.66 | 4039 47 | 4353 40 | 9686 6 | 2.16 | 6312 45 | 6428 42 | 9884 2 |
| 1.67 | 4086 47 | 4394 40 | 9692 6 | 2.17 | 6357 45 | 6470 42 | 9887 2 |
| 1.68 | 4132 47 | 4434 41 | 9698 6 | 2.18 | 6401 45 | 6512 42 | 9889 2 |
| 1.69 | 4179 46 | 4475 41 | 9704 6 | 2.19 | 6446 45 | 6555 42 | 9891 2 |
| **1.70** | 4225 46 | 4515 41 | 9710 6 | **2.20** | 6491 45 | 6597 42 | 9893 2 |
| 1.71 | 4272 46 | 4556 41 | 9716 6 | 2.21 | 6535 44 | 6640 42 | 9895 2 |
| 1.72 | 4318 46 | 4597 41 | 9721 6 | 2.22 | 6580 44 | 6682 42 | 9898 2 |
| 1.73 | 4364 46 | 4637 41 | 9727 5 | 2.23 | 6624 44 | 6724 42 | 9900 2 |
| 1.74 | 4411 46 | 4678 41 | 9732 5 | 2.24 | 6668 44 | 6767 42 | 9902 2 |
| **1.75** | 4457 46 | 4719 41 | 9738 5 | **2.25** | 6713 44 | 6809 42 | 9904 2 |
| 1.76 | 4503 46 | 4760 41 | 9743 5 | 2.26 | 6757 44 | 6852 42 | 9905 2 |
| 1.77 | 4549 46 | 4801 41 | 9748 5 | 2.27 | 6802 44 | 6894 43 | 9907 2 |
| 1.78 | 4595 46 | 4842 41 | 9753 5 | 2.28 | 6846 44 | 6937 43 | 9909 2 |
| 1.79 | 4641 46 | 4883 41 | 9758 5 | 2.29 | 6890 44 | 6979 43 | 9911 2 |
| **1.80** | 4687 46 | 4924 41 | 9763 5 | **2.30** | 6935 44 | 7022 43 | 9913 2 |
| 1.81 | 4733 46 | 4965 41 | 9767 5 | 2.31 | 6979 44 | 7064 43 | 9914 2 |
| 1.82 | 4778 46 | 5006 41 | 9772 5 | 2.32 | 7023 44 | 7107 43 | 9916 2 |
| 1.83 | 4824 46 | 5048 41 | 9776 4 | 2.33 | 7067 44 | 7150 43 | 9918 2 |
| 1.84 | 4870 46 | 5089 41 | 9781 4 | 2.34 | 7112 44 | 7192 43 | 9919 2 |
| **1.85** | 4915 46 | 5130 41 | 9785 4 | **2.35** | 7156 44 | 7235 43 | 9921 2 |
| 1.86 | 4961 46 | 5172 41 | 9789 4 | 2.36 | 7200 44 | 7278 43 | 9923 2 |
| 1.87 | 5007 46 | 5213 41 | 9794 4 | 2.37 | 7244 44 | 7320 43 | 9924 2 |
| 1.88 | 5052 46 | 5254 41 | 9798 4 | 2.38 | 7289 44 | 7363 43 | 9926 1 |
| 1.89 | 5098 45 | 5296 41 | 9802 4 | 2.39 | 7333 44 | 7406 43 | 9927 1 |
| **1.90** | 5143 45 | 5337 42 | 9806 4 | **2.40** | 7377 44 | 7448 43 | 9929 1 |
| 1.91 | 5188 45 | 5379 42 | 9810 4 | 2.41 | 7421 44 | 7491 43 | 9930 1 |
| 1.92 | 5234 45 | 5421 42 | 9813 4 | 2.42 | 7465 44 | 7534 43 | 9931 1 |
| 1.93 | 5279 45 | 5462 42 | 9817 4 | 2.43 | 7509 44 | 7577 43 | 9933 1 |
| 1.94 | 5324 45 | 5504 42 | 9821 4 | 2.44 | 7553 44 | 7619 43 | 9934 1 |
| **1.95** | 5370 45 | 5545 42 | 9824 4 | **2.45** | 7597 44 | 7662 43 | 9935 1 |
| 1.96 | 5415 45 | 5587 42 | 9828 3 | 2.46 | 7642 44 | 7705 43 | 9937 1 |
| 1.97 | 5460 45 | 5629 42 | 9831 3 | 2.47 | 7686 44 | 7748 43 | 9938 1 |
| 1.98 | 5505 45 | 5671 42 | 9834 3 | 2.48 | 7730 44 | 7791 43 | 9939 1 |
| 1.99 | 5550 45 | 5713 42 | 9838 3 | 2.49 | 7774 44 | 7833 43 | 9940 1 |
| **2.00** | 5595 45 | 5754 42 | 9841 3 | **2.50** | 7818 44 | 7876 43 | 9941 1 |

# Logarithms of Hyperbolic Functions.

| x | l Sh x | l Ch x | l Th x | x | l Sh x | l Ch x | l Th x |
|---|---|---|---|---|---|---|---|
| | **0.** | **0.** | **9.** | | | | |
| **2.50** | 7818 44 | 7876 43 | 9941 | **3.0** | 1.0008 436 | 1.0029 432 | 9.9978 |
| 2.51 | 7862 44 | 7919 43 | 9943 | 3.1 | 1.0444 436 | 1.0462 433 | 9.9982 |
| 2.52 | 7906 44 | 7962 43 | 9944 | 3.2 | 1.0880 436 | 1.0894 433 | 9.9986 |
| 2.53 | 7950 44 | 8005 43 | 9945 | 3.3 | 1.1316 435 | 1.1327 433 | 9.9988 |
| 2.54 | 7994 44 | 8048 43 | 9946 | 3.4 | 1.1751 435 | 1.1761 433 | 9.9990 |
| **2.55** | 8038 44 | 8091 43 | 9947 | **3.5** | 1.2186 435 | 1.2194 434 | 9.9992 |
| 2.56 | 8082 44 | 8134 43 | 9948 | 3.6 | 1.2621 435 | 1.2628 434 | 9.9994 |
| 2.57 | 8126 44 | 8176 43 | 9949 | 3.7 | 1.3056 435 | 1.3061 434 | 9.9995 |
| 2.58 | 8169 44 | 8219 43 | 9950 | 3.8 | 1.3491 435 | 1.3495 434 | 9.9996 |
| 2.59 | 8213 44 | 8262 43 | 9951 | 3.9 | 1.3925 435 | 1.3929 434 | 9.9996 |
| **2.60** | 8257 44 | 8305 43 | 9952 | **4.0** | 1.4360 435 | 1.4363 434 | 9.9997 |
| 2.61 | 8301 44 | 8348 43 | 9953 | 4.1 | 1.4795 435 | 1.4797 434 | 9.9998 |
| 2.62 | 8345 44 | 8391 43 | 9954 | 4.2 | 1.5229 434 | 1.5231 434 | 9.9998 |
| 2.63 | 8389 44 | 8434 43 | 9955 | 4.3 | 1.5664 434 | 1.5665 434 | 9.9998 |
| 2.64 | 8433 44 | 8477 43 | 9956 | 4.4 | 1.6098 434 | 1.6099 434 | 9.9999 |
| **2.65** | 8477 44 | 8520 43 | 9957 | **4.5** | 1.6532 434 | 1.6533 434 | 9.9999 |
| 2.66 | 8521 44 | 8563 43 | 9958 | 4.6 | 1.6967 434 | 1.6968 434 | 9.9999 |
| 2.67 | 8564 44 | 8606 43 | 9958 | 4.7 | 1.7401 434 | 1.7402 434 | 9.9999 |
| 2.68 | 8608 44 | 8649 43 | 9959 | 4.8 | 1.7836 434 | 1.7836 434 | 9.9999 |
| 2.69 | 8652 44 | 8692 43 | 9960 | 4.9 | 1.8270 434 | 1.8270 434 | 0.0000 |
| **2.70** | 8696 44 | 8735 43 | 9961 | **5.0** | 1.8704 434 | 1.8705 434 | 0.0000 |
| 2.71 | 8740 44 | 8778 43 | 9962 | 5.1 | 1.9139 434 | 1.9139 434 | 0.0000 |
| 2.72 | 8784 44 | 8821 43 | 9962 | 5.2 | 1.9573 434 | 1.9573 434 | 0.0000 |
| 2.73 | 8827 44 | 8864 43 | 9963 | 5.3 | 2.0007 434 | 2.0007 434 | 0.0000 |
| 2.74 | 8871 44 | 8907 43 | 9964 | 5.4 | 2.0442 434 | 2.0442 434 | 0.0000 |
| **2.75** | 8915 44 | 8951 43 | 9965 | **5.5** | 2.0876 434 | 2.0876 434 | 0.0000 |
| 2.76 | 8959 44 | 8994 43 | 9965 | 5.6 | 2.1310 434 | 2.1310 434 | 0.0000 |
| 2.77 | 9003 44 | 9037 43 | 9966 | 5.7 | 2.1744 434 | 2.1745 434 | 0.0000 |
| 2.78 | 9046 44 | 9080 43 | 9967 | 5.8 | 2.2179 434 | 2.2179 434 | 0.0000 |
| 2.79 | 9090 44 | 9123 43 | 9967 | 5.9 | 2.2613 434 | 2.2613 434 | 0.0000 |
| **2.80** | 9134 44 | 9166 43 | 9968 | **6.0** | 2.3047 4343 | 2.3047 4343 | 0.0000 |
| 2.81 | 9178 44 | 9209 43 | 9969 | 7.0 | 2.7390 4343 | 2.7390 4343 | 0.0000 |
| 2.82 | 9221 44 | 9252 43 | 9969 | 8.0 | 3.1733 4343 | 3.1733 4343 | 0.0000 |
| 2.83 | 9265 44 | 9295 43 | 9970 | 9.0 | 3.6076 4343 | 3.6076 4343 | 0.0000 |
| 2.84 | 9309 44 | 9338 43 | 9970 | **10.0** | 4.0419 4343 | 4.0419 4343 | 0.0000 |
| **2.85** | 9353 44 | 9382 43 | 9971 | | | | |
| 2.86 | 9396 44 | 9425 43 | 9972 | | | | |
| 2.87 | 9440 44 | 9468 43 | 9972 | | | | |
| 2.88 | 9484 44 | 9511 43 | 9973 | | | | |
| 2.89 | 9527 44 | 9554 43 | 9973 | | | | |
| **2.90** | 9571 44 | 9597 43 | 9974 | | | | |
| 2.91 | 9615 44 | 9641 43 | 9974 | | | | |
| 2.92 | 9658 44 | 9684 43 | 9975 | | | | |
| 2.93 | 9702 44 | 9727 43 | 9975 | | | | |
| 2.94 | 9746 44 | 9770 43 | 9976 | | | | |
| **2.95** | 9789 44 | 9813 43 | 9976 | | | | |
| 2.96 | 9833 44 | 9856 43 | 9977 | | | | |
| 2.97 | 9877 44 | 9900 43 | 9977 | | | | |
| 2.98 | 9920 44 | 9943 43 | 9978 | | | | |
| 2.99 | 9964 44 | 9986 43 | 9978 | | | | |
| **3.00** | 1.0008 44 | 1.0029 43 | 9978 | | | | |

For higher values:

$$\log \text{Sh } x = \log \text{Ch } x = x\mu - 0.301030;$$
$$\text{Sh}^{-1} u = \text{Ch}^{-1} u = (\log u + 0.3010)\, \mu^{-1}.$$

| $n$ | $n\mu$ | $n\mu^{-1}$ | $n$ |
|---|---|---|---|
| 1 | 0.434294 | 2.302585 | 1 |
| 2 | 0.868589 | 4.605170 | 2 |
| 3 | 1.302883 | 6.907755 | 3 |
| 4 | 1.737178 | 9.210340 | 4 |
| 5 | 2.171472 | 11.512925 | 5 |
| 6 | 2.605767 | 13.815511 | 6 |
| 7 | 3.040061 | 16.118096 | 7 |
| 8 | 3.474356 | 18.420681 | 8 |
| 9 | 3.908650 | 20.723266 | 9 |
| 10 | 4.342945 | 23.025851 | 10 |

# Natural Sines and Cosines.

| φ | sin φ | cos φ | |
|---|---|---|---|
| 0° | .000000 | 1.0000 | 90° |
| 10' | .002909 | 1.0000 | 50' |
| 20' | .005818 | 1.0000 | 40' |
| 30' | .008727 | 1.0000 | 30' |
| 40' | .011635 | 0.9999 | 20' |
| 50' | .014544 | 0.9999 | 10' |
| 1° | .017452 | 0.9998 | 89° |
| 10' | .02036 | 0.9998 | 50' |
| 20' | .02327 | 0.9997 | 40' |
| 30' | .02618 | 0.9997 | 30' |
| 40' | .02908 | 0.9996 | 20' |
| 50' | .03199 | 0.9995 | 10' |
| 2° | .03490 | 0.9994 | 88° |
| 10' | .03781 | 0.9993 | 50' |
| 20' | .04071 | 0.9992 | 40' |
| 30' | .04362 | 0.9990 | 30' |
| 40' | .04653 | 0.9989 | 20' |
| 50' | .04943 | 0.9988 | 10' |
| 3° | .05234 | 0.9986 | 87° |
| 10' | .05524 | 0.9985 | 50' |
| 20' | .05814 | 0.9983 | 40' |
| 30' | .06105 | 0.9981 | 30' |
| 40' | .06395 | 0.9980 | 20' |
| 50' | .06685 | 0.9978 | 10' |
| 4° | .06976 | 0.9976 | 86° |
| 10' | .07266 | 0.9974 | 50' |
| 20' | .07556 | 0.9971 | 40' |
| 30' | .07846 | 0.9969 | 30' |
| 40' | .08136 | 0.9967 | 20' |
| 50' | .08426 | 0.9964 | 10' |
| 5° | .08716 | 0.9962 | 85° |
| 10' | .09005 | 0.9959 | 50' |
| 20' | .09295 | 0.9957 | 40' |
| 30' | .09585 | 0.9954 | 30' |
| 40' | .09874 | 0.9951 | 20' |
| 50' | .10164 | 0.9948 | 10' |
| 6° | .10453 | 0.9945 | 84° |
| 10' | .1074 | 0.9942 | 50' |
| 20' | .1103 | 0.9939 | 40' |
| 30' | .1132 | 0.9936 | 30' |
| 40' | .1161 | 0.9932 | 20' |
| 50' | .1190 | 0.9929 | 10' |
| 7° | .1219 | 0.9925 | 83° |
| 10' | .1248 | 0.9922 | 50' |
| 20' | .1276 | 0.9918 | 40' |
| 30' | .1305 | 0.9914 | 30' |
| | cos θ | sin θ | θ |

| φ | sin φ | cos φ | |
|---|---|---|---|
| 7° | | | |
| 30' | .1305 | .9914 | 30' |
| 40' | .1334 | .9911 | 20' |
| 50' | .1363 | .9907 | 10' |
| 8° | .1392 | .9903 | 82° |
| 10' | .1421 | .9899 | 50' |
| 20' | .1449 | .9894 | 40' |
| 30' | .1478 | .9890 | 30' |
| 40' | .1507 | .9886 | 20' |
| 50' | .1536 | .9881 | 10' |
| 9° | .1564 | .9877 | 81° |
| 10' | .1593 | .9872 | 50' |
| 20' | .1622 | .9868 | 40' |
| 30' | .1650 | .9863 | 30' |
| 40' | .1679 | .9858 | 20' |
| 50' | .1708 | .9853 | 10' |
| 10° | .1736 | .9848 | 80° |
| 10' | .1765 | .9843 | 50' |
| 20' | .1794 | .9838 | 40' |
| 30' | .1822 | .9833 | 30' |
| 40' | .1851 | .9827 | 20' |
| 50' | .1880 | .9822 | 10' |
| 11° | .1908 | .9816 | 79° |
| 10' | .1937 | .9811 | 50' |
| 20' | .1965 | .9805 | 40' |
| 30' | .1994 | .9799 | 30' |
| 40' | .2022 | .9793 | 20' |
| 50' | .2051 | .9787 | 10' |
| 12° | .2079 | .9781 | 78° |
| 10' | .2108 | .9775 | 50' |
| 20' | .2136 | .9769 | 40' |
| 30' | .2164 | .9763 | 30' |
| 40' | .2193 | .9757 | 20' |
| 50' | .2221 | .9750 | 10' |
| 13° | .2250 | .9744 | 77° |
| 10' | .2278 | .9737 | 50' |
| 20' | .2306 | .9730 | 40' |
| 30' | .2334 | .9724 | 30' |
| 40' | .2363 | .9717 | 20' |
| 50' | .2391 | .9710 | 10' |
| 14° | .2419 | .9703 | 76° |
| 10' | .2447 | .9696 | 50' |
| 20' | .2476 | .9689 | 40' |
| 30' | .2504 | .9681 | 30' |
| 40' | .2532 | .9674 | 20' |
| 50' | .2560 | .9667 | 10' |
| 15° | .2588 | .9659 | 75° |
| | cos θ | sin θ | θ |

| φ | sin φ | cos φ | |
|---|---|---|---|
| 15° | .2588 | .9659 | 75° |
| 10' | .2616 | .9652 | 50' |
| 20' | .2644 | .9644 | 40' |
| 30' | .2672 | .9636 | 30' |
| 40' | .2700 | .9628 | 20' |
| 50' | .2728 | .9621 | 10' |
| 16° | .2756 | .9613 | 74° |
| 10' | .2784 | .9605 | 50' |
| 20' | .2812 | .9596 | 40' |
| 30' | .2840 | .9588 | 30' |
| 40' | .2868 | .9580 | 20' |
| 50' | .2896 | .9572 | 10' |
| 17° | .2924 | .9563 | 73° |
| 10' | .2952 | .9555 | 50' |
| 20' | .2979 | .9546 | 40' |
| 30' | .3007 | .9537 | 30' |
| 40' | .3035 | .9528 | 20' |
| 50' | .3062 | .9520 | 10' |
| 18° | .3090 | .9511 | 72° |
| 10' | .3118 | .9502 | 50' |
| 20' | .3145 | .9492 | 40' |
| 30' | .3173 | .9483 | 30' |
| 40' | .3201 | .9474 | 20' |
| 50' | .3228 | .9465 | 10' |
| 19° | .3256 | .9455 | 71° |
| 10' | .3283 | .9446 | 50' |
| 20' | .3311 | .9436 | 40' |
| 30' | .3338 | .9426 | 30' |
| 40' | .3365 | .9417 | 20' |
| 50' | .3393 | .9407 | 10' |
| 20° | .3420 | .9397 | 70° |
| 10' | .3448 | .9387 | 50' |
| 20' | .3475 | .9377 | 40' |
| 30' | .3502 | .9367 | 30' |
| 40' | .3529 | .9356 | 20' |
| 50' | .3557 | .9346 | 10' |
| 21° | .3584 | .9336 | 69° |
| 10' | .3611 | .9325 | 50' |
| 20' | .3638 | .9315 | 40' |
| 30' | .3665 | .9304 | 30' |
| 40' | .3692 | .9293 | 20' |
| 50' | .3719 | .9283 | 10' |
| 22° | .3746 | .9272 | 68° |
| 10' | .3773 | .9261 | 50' |
| 20' | .3800 | .9250 | 40' |
| 30' | .3827 | .9239 | 30' |
| | | | 67° |
| | cos θ | sin θ | θ |

# Natural Sines and Cosines.

| φ | sin φ | cos φ | |
|---|---|---|---|
| **22°** | | | 30' |
| 30' | .3827 | .9239 | 30' |
| 40' | .3854 | .9228 | 20' |
| 50' | .3881 | .9216 | 10' |
| **23°** | .3907 | .9205 | **67°** |
| 10' | .3934 | .9194 | 50' |
| 20' | .3961 | .9182 | 40' |
| 30' | .3987 | .9171 | 30' |
| 40' | .4014 | .9159 | 20' |
| 50' | .4041 | .9147 | 10' |
| **24°** | .4067 | .9135 | **66°** |
| 10' | .4094 | .9124 | 50' |
| 20' | .4120 | .9112 | 40' |
| 30' | .4147 | .9100 | 30' |
| 40' | .4173 | .9088 | 20' |
| 50' | .4200 | .9075 | 10' |
| **25°** | .4226 | .9063 | **65°** |
| 10' | .4253 | .9051 | 50' |
| 20' | .4279 | .9038 | 40' |
| 30' | .4305 | .9026 | 30' |
| 40' | .4331 | .9013 | 20' |
| 50' | .4358 | .9001 | 10' |
| **26°** | .4384 | .8988 | **64°** |
| 10' | .4410 | .8975 | 50' |
| 20' | .4436 | .8962 | 40' |
| 30' | .4462 | .8949 | 30' |
| 40' | .4488 | .8936 | 20' |
| 50' | .4514 | .8923 | 10' |
| **27°** | .4540 | .8910 | **63°** |
| 10' | .4566 | .8897 | 50' |
| 20' | .4592 | .8884 | 40' |
| 30' | .4617 | .8870 | 30' |
| 40' | .4643 | .8857 | 20' |
| 50' | .4669 | .8843 | 10' |
| **28°** | .4695 | .8829 | **62°** |
| 10' | .4720 | .8816 | 50' |
| 20' | .4746 | .8802 | 40' |
| 30' | .4772 | .8788 | 30' |
| 40' | .4797 | .8774 | 20' |
| 50' | .4823 | .8760 | 10' |
| **29°** | .4848 | .8746 | **61°** |
| 10' | .4874 | .8732 | 50' |
| 20' | .4899 | .8718 | 40' |

| φ | sin φ | cos φ | |
|---|---|---|---|
| **30°** | .5000 | .8660 | **60°** |
| 10' | .5025 | .8646 | 50' |
| 20' | .5050 | .8631 | 40' |
| 30' | .5075 | .8616 | 30' |
| 40' | .5100 | .8601 | 20' |
| 50' | .5125 | .8587 | 10' |
| **31°** | .5150 | .8572 | **59°** |
| 10' | .5175 | .8557 | 50' |
| 20' | .5200 | .8542 | 40' |
| 30' | .5225 | .8526 | 30' |
| 40' | .5250 | .8511 | 20' |
| 50' | .5275 | .8496 | 10' |
| **32°** | .5299 | .8480 | **58°** |
| 10' | .5324 | .8465 | 50' |
| 20' | .5348 | .8450 | 40' |
| 30' | .5373 | .8434 | 30' |
| 40' | .5398 | .8418 | 20' |
| 50' | .5422 | .8403 | 10' |
| **33°** | .5446 | .8387 | **57°** |
| 10' | .5471 | .8371 | 50' |
| 20' | .5495 | .8355 | 40' |
| 30' | .5519 | .8339 | 30' |
| 40' | .5544 | .8323 | 20' |
| 50' | .5568 | .8307 | 10' |
| **34°** | .5592 | .8290 | **56°** |
| 10' | .5616 | .8274 | 50' |
| 20' | .5640 | .8258 | 40' |
| 30' | .5664 | .8241 | 30' |
| 40' | .5688 | .8225 | 20' |
| 50' | .5712 | .8208 | 10' |
| **35°** | .5736 | .8192 | **55°** |
| 10' | .5760 | .8175 | 50' |
| 20' | .5783 | .8158 | 40' |
| 30' | .5807 | .8141 | 30' |
| 40' | .5831 | .8124 | 20' |
| 50' | .5854 | .8107 | 10' |
| **36°** | .5878 | .8090 | **54°** |
| 10' | .5901 | .8073 | 50' |
| 20' | .5925 | .8056 | 40' |
| 30' | .5948 | .8039 | 30' |
| 40' | .5972 | .8021 | 20' |
| 50' | .5995 | .8004 | 10' |
| **37°** | .6018 | .7986 | **53°** |
| 20' | | | 40' |

| φ | sin φ | cos φ | |
|---|---|---|---|
| **37°** | | | |
| 30' | .6088 | .7934 | 30' |
| 40' | .6111 | .7916 | 20' |
| 50' | .6134 | .7898 | 10' |
| **38°** | .6157 | .7880 | **52°** |
| 10' | .6180 | .7862 | 50' |
| 20' | .6202 | .7844 | 40' |
| 30' | .6225 | .7826 | 30' |
| 40' | .6248 | .7808 | 20' |
| 50' | .6271 | .7790 | 10' |
| **39°** | .6293 | .7771 | **51°** |
| 10' | .6316 | .7753 | 50' |
| 20' | .6338 | .7735 | 40' |
| 30' | .6361 | .7716 | 30' |
| 40' | .6383 | .7698 | 20' |
| 50' | .6406 | .7679 | 10' |
| **40°** | .6428 | .7660 | **50°** |
| 10' | .6450 | .7642 | 50' |
| 20' | .6472 | .7623 | 40' |
| 30' | .6494 | .7604 | 30' |
| 40' | .6517 | .7585 | 20' |
| 50' | .6539 | .7566 | 10' |
| **41°** | .6561 | .7547 | **49°** |
| 10' | .6583 | .7528 | 50' |
| 20' | .6604 | .7509 | 40' |
| 30' | .6626 | .7490 | 30' |
| 40' | .6648 | .7470 | 20' |
| 50' | .6670 | .7451 | 10' |
| **42°** | .6691 | .7431 | **48°** |
| 10' | .6713 | .7412 | 50' |
| 20' | .6734 | .7392 | 40' |
| 30' | .6756 | .7373 | 30' |
| 40' | .6777 | .7353 | 20' |
| 50' | .6799 | .7333 | 10' |
| **43°** | .6820 | .7314 | **47°** |
| 10' | .6841 | .7294 | 50' |
| 20' | .6862 | .7274 | 40' |
| 30' | .6884 | .7254 | 30' |
| 40' | .6905 | .7234 | 20' |
| 50' | .6926 | .7214 | 10' |
| **44°** | .6947 | .7193 | **46°** |
| 10' | .6967 | .7173 | 50' |
| 20' | .6988 | .7153 | 40' |

# Natural Tangents and Cotangents.

| φ | tan φ | ctn φ | θ |
|---|---|---|---|
| 0° | .000000 | | 90° |
| 10' | .002909 | | 50' |
| 20' | .005818 | | 40' |
| 30' | .008727 | | 30' |
| 40' | .011636 | | 20' |
| 50' | .014545 | | 10' |
| 1° | .017455 | 57. | 89° |
| 10' | .02036 | 49. | 50' |
| 20' | .02328 | 43. | 40' |
| 30' | .02619 | 38. | 30' |
| 40' | .02910 | 34. | 20' |
| 50' | .03201 | 31. | 10' |
| 2° | .03492 | 28.6 | 88° |
| 10' | .03783 | 26.4 | 50' |
| 20' | .04075 | 24.5 | 40' |
| 30' | .04366 | 22.9 | 30' |
| 40' | .04658 | 21.5 | 20' |
| 50' | .04949 | 20.2 | 10' |
| 3° | .05241 | 19.1 | 87° |
| 10' | .05533 | 18.1 | 50' |
| 20' | .05824 | 17.2 | 40' |
| 30' | .06116 | 16.3 | 30' |
| 40' | .06408 | 15.6 | 20' |
| 50' | .06700 | 14.9 | 10' |
| 4° | .06993 | 14.3 | 86° |
| 10' | .07285 | 13.73 | 50' |
| 20' | .07578 | 13.20 | 40' |
| 30' | .07870 | 12.71 | 30' |
| 40' | .08163 | 12.25 | 20' |
| 50' | .08456 | 11.83 | 10' |
| 5° | .08749 | 11.43 | 85° |
| 10' | .09042 | 11.06 | 50' |
| 20' | .09335 | 10.71 | 40' |
| 30' | .09629 | 10.39 | 30' |
| 40' | .09923 | 10.08 | 20' |
| 50' | .10216 | 9.79 | 10' |
| 6° | .10510 | 9.51 | 84° |
| 10' | .1080 | 9.26 | 50' |
| 20' | .1110 | 9.01 | 40' |
| 30' | .1139 | 8.78 | 30' |
| 40' | .1169 | 8.56 | 20' |
| 50' | .1198 | 8.34 | 10' |
| 7° | .1228 | 8.14 | 83° |
| 10' | .1257 | 7.95 | 50' |
| 20' | .1287 | 7.77 | 40' |
| 30' | .1317 | 7.60 | 30' |
| | ctn θ | tan θ | θ |

| φ | tan φ | ctn φ | θ |
|---|---|---|---|
| 7° | | | |
| 30' | .1317 | 7.60 | 30' |
| 40' | .1346 | 7.43 | 20' |
| 50' | .1376 | 7.27 | 10' |
| 8° | .1405 | 7.12 | 82° |
| 10' | .1435 | 6.97 | 50' |
| 20' | .1465 | 6.83 | 40' |
| 30' | .1495 | 6.69 | 30' |
| 40' | .1524 | 6.56 | 20' |
| 50' | .1554 | 6.43 | 10' |
| 9° | .1584 | 6.31 | 81° |
| 10' | .1614 | 6.197 | 50' |
| 20' | .1644 | 6.084 | 40' |
| 30' | .1673 | 5.976 | 30' |
| 40' | .1703 | 5.871 | 20' |
| 50' | .1733 | 5.769 | 10' |
| 10° | .1763 | 5.671 | 80° |
| 10' | .1793 | 5.576 | 50' |
| 20' | .1823 | 5.485 | 40' |
| 30' | .1853 | 5.396 | 30' |
| 40' | .1883 | 5.309 | 20' |
| 50' | .1914 | 5.226 | 10' |
| 11° | .1944 | 5.145 | 79° |
| 10' | .1974 | 5.066 | 50' |
| 20' | .2004 | 4.989 | 40' |
| 30' | .2035 | 4.915 | 30' |
| 40' | .2065 | 4.843 | 20' |
| 50' | .2095 | 4.773 | 10' |
| 12° | .2126 | 4.705 | 78° |
| 10' | .2156 | 4.638 | 50' |
| 20' | .2186 | 4.574 | 40' |
| 30' | .2217 | 4.511 | 30' |
| 40' | .2247 | 4.449 | 20' |
| 50' | .2278 | 4.390 | 10' |
| 13° | .2309 | 4.331 | 77° |
| 10' | .2339 | 4.275 | 50' |
| 20' | .2370 | 4.219 | 40' |
| 30' | .2401 | 4.165 | 30' |
| 40' | .2432 | 4.113 | 20' |
| 50' | .2462 | 4.061 | 10' |
| 14° | .2493 | 4.011 | 76° |
| 10' | .2524 | 3.962 | 50' |
| 20' | .2555 | 3.914 | 40' |
| 30' | .2586 | 3.867 | 30' |
| 40' | .2617 | 3.821 | 20' |
| 50' | .2648 | 3.776 | 10' |
| 15° | .2679 | 3.732 | 75° |
| | ctn θ | tan θ | θ |

| φ | tan φ | ctn φ | θ |
|---|---|---|---|
| 15° | .2679 | 3.732 | 75° |
| 10' | .2711 | 3.689 | 50' |
| 20' | .2742 | 3.647 | 40' |
| 30' | .2773 | 3.606 | 30' |
| 40' | .2805 | 3.566 | 20' |
| 50' | .2836 | 3.526 | 10' |
| 16° | .2867 | 3.487 | 74° |
| 10' | .2899 | 3.450 | 50' |
| 20' | .2931 | 3.412 | 40' |
| 30' | .2962 | 3.376 | 30' |
| 40' | .2994 | 3.340 | 20' |
| 50' | .3026 | 3.305 | 10' |
| 17° | .3057 | 3.271 | 73° |
| 10' | .3089 | 3.237 | 50' |
| 20' | .3121 | 3.204 | 40' |
| 30' | .3153 | 3.172 | 30' |
| 40' | .3185 | 3.140 | 20' |
| 50' | .3217 | 3.108 | 10' |
| 18° | .3249 | 3.078 | 72° |
| 10' | .3281 | 3.047 | 50' |
| 20' | .3314 | 3.018 | 40' |
| 30' | .3346 | 2.989 | 30' |
| 40' | .3378 | 2.960 | 20' |
| 50' | .3411 | 2.932 | 10' |
| 19° | .3443 | 2.904 | 71° |
| 10' | .3476 | 2.877 | 50' |
| 20' | .3508 | 2.850 | 40' |
| 30' | .3541 | 2.824 | 30' |
| 40' | .3574 | 2.798 | 20' |
| 50' | .3607 | 2.773 | 10' |
| 20° | .3640 | 2.747 | 70° |
| 10' | .3673 | 2.723 | 50' |
| 20' | .3706 | 2.699 | 40' |
| 30' | .3739 | 2.675 | 30' |
| 40' | .3772 | 2.651 | 20' |
| 50' | .3805 | 2.628 | 10' |
| 21° | .3839 | 2.605 | 69° |
| 10' | .3872 | 2.583 | 50' |
| 20' | .3906 | 2.560 | 40' |
| 30' | .3939 | 2.539 | 30' |
| 40' | .3973 | 2.517 | 20' |
| 50' | .4006 | 2.496 | 10' |
| 22° | .4040 | 2.475 | 68° |
| 10' | .4074 | 2.455 | 50' |
| 20' | .4108 | 2.434 | 40' |
| 30' | .4142 | 2.414 | 30' |
| | | | 67° |
| | ctn θ | tan θ | θ |

# Natural Tangents and Cotangents.

| φ | tan φ | ctn φ | | φ | tan φ | ctn φ | | φ | tan φ | ctn φ | |
|---|---|---|---|---|---|---|---|---|---|---|---|
| 22° | | | | 30° | .5774 | 1.732 | 60° | 37° | | | |
| 30' | .4142 | 2.414 | 30' | 10' | .5812 | 1.720 | 50' | 30' | .7673 | 1.303 | 30' |
| 40' | .4176 | 2.394 | 20' | 20' | .5851 | 1.709 | 40' | 40' | .7720 | 1.295 | 20' |
| 50' | .4210 | 2.375 | 10' | 30' | .5890 | 1.698 | 30' | 50' | .7766 | 1.288 | 10' |
| 23° | .4245 | 2.356 | 67° | 40' | .5930 | 1.686 | 20' | 38° | .7813 | 1.280 | 52° |
| 10' | .4279 | 2.337 | 50' | 50' | .5969 | 1.675 | 10' | 10' | .7860 | 1.272 | 50' |
| 20' | .4314 | 2.318 | 40' | 31° | .6009 | 1.664 | 59° | 20' | .7907 | 1.265 | 40' |
| 30' | .4348 | 2.300 | 30' | 10' | .6048 | 1.653 | 50' | 30' | .7954 | 1.257 | 30' |
| 40' | .4383 | 2.282 | 20' | 20' | .6088 | 1.643 | 40' | 40' | .8002 | 1.250 | 20' |
| 50' | .4417 | 2.264 | 10' | 30' | .6128 | 1.632 | 30' | 50' | .8050 | 1.242 | 10' |
| 24° | .4452 | 2.246 | 66° | 40' | .6168 | 1.621 | 20' | 39° | .8098 | 1.235 | 51° |
| 10' | .4487 | 2.229 | 50' | 50' | .6208 | 1.611 | 10' | 10' | .8146 | 1.228 | 50' |
| 20' | .4522 | 2.211 | 40' | 32° | .6249 | 1.600 | 58° | 20' | .8195 | 1.220 | 40' |
| 30' | .4557 | 2.194 | 30' | 10' | .6289 | 1.590 | 50' | 30' | .8243 | 1.213 | 30' |
| 40' | .4592 | 2.177 | 20' | 20' | .6330 | 1.580 | 40' | 40' | .8292 | 1.206 | 20' |
| 50' | .4628 | 2.161 | 10' | 30' | .6371 | 1.570 | 30' | 50' | .8342 | 1.199 | 10' |
| 25° | .4663 | 2.145 | 65° | 40' | .6412 | 1.560 | 20' | 40° | .8391 | 1.192 | 50° |
| 10' | .4699 | 2.128 | 50' | 50' | .6453 | 1.550 | 10' | 10' | .8441 | 1.185 | 50' |
| 20' | .4734 | 2.112 | 40' | 33° | .6494 | 1.540 | 57° | 20' | .8491 | 1.178 | 40' |
| 30' | .4770 | 2.097 | 30' | 10' | .6536 | 1.530 | 50' | 30' | .8541 | 1.171 | 30' |
| 40' | .4806 | 2.081 | 20' | 20' | .6577 | 1.520 | 40' | 40' | .8591 | 1.164 | 20' |
| 50' | .4841 | 2.066 | 10' | 30' | .6619 | 1.511 | 30' | 50' | .8642 | 1.157 | 10' |
| 26° | .4877 | 2.050 | 64° | 40' | .6661 | 1.501 | 20' | 41° | .8693 | 1.150 | 49° |
| 10' | .4913 | 2.035 | 50' | 50' | .6703 | 1.492 | 10' | 10' | .8744 | 1.144 | 50' |
| 20' | .4950 | 2.020 | 40' | 34° | .6745 | 1.483 | 56° | 20' | .8796 | 1.137 | 40' |
| 30' | .4986 | 2.006 | 30' | 10' | .6787 | 1.473 | 50' | 30' | .8847 | 1.130 | 30' |
| 40' | .5022 | 1.991 | 20' | 20' | .6830 | 1.464 | 40' | 40' | .8899 | 1.124 | 20' |
| 50' | .5059 | 1.977 | 10' | 30' | .6873 | 1.455 | 30' | 50' | .8952 | 1.117 | 10' |
| 27° | .5095 | 1.963 | 63° | 40' | .6916 | 1.446 | 20' | 42° | .9004 | 1.111 | 48° |
| 10' | .5132 | 1.949 | 50' | 50' | .6959 | 1.437 | 10' | 10' | .9057 | 1.104 | 50' |
| 20' | .5169 | 1.935 | 40' | 35° | .7002 | 1.428 | 55° | 20' | .9110 | 1.098 | 40' |
| 30' | .5206 | 1.921 | 30' | 10' | .7046 | 1.419 | 50' | 30' | .9163 | 1.091 | 30' |
| 40' | .5243 | 1.907 | 20' | 20' | .7089 | 1.411 | 40' | 40' | .9217 | 1.085 | 20' |
| 50' | .5280 | 1.894 | 10' | 30' | .7133 | 1.402 | 30' | 50' | .9271 | 1.079 | 10' |
| 28° | .5317 | 1.881 | 62° | 40' | .7177 | 1.393 | 20' | 43° | .9325 | 1.072 | 47° |
| 10' | .5354 | 1.868 | 50' | 50' | .7221 | 1.385 | 10' | 10' | .9380 | 1.066 | 50' |
| 20' | .5392 | 1.855 | 40' | 36° | .7265 | 1.376 | 54° | 20' | .9435 | 1.060 | 40' |
| 30' | .5430 | 1.842 | 30' | 10' | .7310 | 1.368 | 50' | 30' | .9490 | 1.054 | 30' |
| 40' | .5467 | 1.829 | 20' | 20' | .7355 | 1.360 | 40' | 40' | .9545 | 1.048 | 20' |
| 50' | .5505 | 1.816 | 10' | 30' | .7400 | 1.351 | 30' | 50' | .9601 | 1.042 | 10' |
| 29° | .5543 | 1.804 | 61° | 40' | .7445 | 1.343 | 20' | 44° | .9657 | 1.036 | 46° |
| 10' | .5581 | 1.792 | 50' | 50' | .7490 | 1.335 | 10' | 10' | .9713 | 1.030 | 50' |
| 20' | .5619 | 1.780 | 40' | 37° | .7536 | 1.327 | 53° | 20' | .9770 | 1.024 | 40' |
| 30' | .5658 | 1.767 | 30' | 10' | .7581 | 1.319 | 50' | 30' | .9827 | 1.018 | 30' |
| 40' | .5696 | 1.756 | 20' | 20' | .7627 | 1.311 | 40' | 40' | .9884 | 1.012 | 20' |
| 50' | .5735 | 1.744 | 10' | 30' | .7673 | 1.303 | 30' | 50' | .9942 | 1.006 | 10' |
| 30° | .5774 | 1.732 | 60° | | | | 52° | 45° | 1.0000 | 1.000 | 45° |
| | ctn θ | tan θ | θ | | ctn θ | tan θ | θ | | ctn θ | tan θ | θ |

# Natural Secants and Cosecants.

| φ | sec φ | csc φ | |
|---|---|---|---|
| 0° | 1.0000 | | 90° |
| 10' | 1.0000 | | 50' |
| 20' | 1.0000 | | 40' |
| 30' | 1.0000 | | 30' |
| 40' | 1.0001 | | 20' |
| 50' | 1.0001 | | 10' |
| 1° | 1.0002 | 57. | 89° |
| 10' | 1.0002 | 49. | 50' |
| 20' | 1.0003 | 43. | 40' |
| 30' | 1.0003 | 38. | 30' |
| 40' | 1.0004 | 34. | 20' |
| 50' | 1.0005 | 31. | 10' |
| 2° | 1.0006 | 28.7 | 88° |
| 10' | 1.0007 | 26.5 | 50' |
| 20' | 1.0008 | 24.6 | 40' |
| 30' | 1.0010 | 22.9 | 30' |
| 40' | 1.0011 | 21.5 | 20' |
| 50' | 1.0012 | 20.2 | 10' |
| 3° | 1.0014 | 19.1 | 87° |
| 10' | 1.0015 | 18.1 | 50' |
| 20' | 1.0017 | 17.2 | 40' |
| 30' | 1.0019 | 16.4 | 30' |
| 40' | 1.0021 | 15.6 | 20' |
| 50' | 1.0022 | 15.0 | 10' |
| 4° | 1.0024 | 14.3 | 86° |
| 10' | 1.0026 | 13.76 | 50' |
| 20' | 1.0029 | 13.23 | 40' |
| 30' | 1.0031 | 12.75 | 30' |
| 40' | 1.0033 | 12.29 | 20' |
| 50' | 1.0036 | 11.87 | 10' |
| 5° | 1.0038 | 11.47 | 85° |
| 10' | 1.0041 | 11.10 | 50' |
| 20' | 1.0043 | 10.76 | 40' |
| 30' | 1.0046 | 10.43 | 30' |
| 40' | 1.0049 | 10.13 | 20' |
| 50' | 1.0052 | 9.84 | 10' |
| 6° | 1.0055 | 9.57 | 84° |
| 10' | 1.0058 | 9.31 | 50' |
| 20' | 1.0061 | 9.07 | 40' |
| 30' | 1.0065 | 8.83 | 30' |
| 40' | 1.0068 | 8.61 | 20' |
| 50' | 1.0072 | 8.40 | 10' |
| 7° | 1.0075 | 8.21 | 83° |
| 10' | 1.0079 | 8.02 | 50' |
| 20' | 1.0082 | 7.83 | 40' |
| 30' | 1.0086 | 7.66 | 30' |
| | | | 82° |
| | csc θ | sec θ | θ |

| φ | sec φ | csc φ | |
|---|---|---|---|
| 7° | | | |
| 30' | 1.009 | 7.66 | 30' |
| 40' | 1.009 | 7.50 | 20' |
| 50' | 1.009 | 7.34 | 10' |
| 8° | 1.010 | 7.19 | 82° |
| 10' | 1.010 | 7.04 | 50' |
| 20' | 1.011 | 6.90 | 40' |
| 30' | 1.011 | 6.77 | 30' |
| 40' | 1.012 | 6.64 | 20' |
| 50' | 1.012 | 6.51 | 10' |
| 9° | 1.012 | 6.39 | 81° |
| 10' | 1.013 | 6.277 | 50' |
| 20' | 1.013 | 6.166 | 40' |
| 30' | 1.014 | 6.059 | 30' |
| 40' | 1.014 | 5.955 | 20' |
| 50' | 1.015 | 5.855 | 10' |
| 10° | 1.015 | 5.759 | 80° |
| 10' | 1.016 | 5.665 | 50' |
| 20' | 1.016 | 5.575 | 40' |
| 30' | 1.017 | 5.487 | 30' |
| 40' | 1.018 | 5.403 | 20' |
| 50' | 1.018 | 5.320 | 10' |
| 11° | 1.019 | 5.241 | 79° |
| 10' | 1.019 | 5.164 | 50' |
| 20' | 1.020 | 5.089 | 40' |
| 30' | 1.020 | 5.016 | 30' |
| 40' | 1.021 | 4.945 | 20' |
| 50' | 1.022 | 4.876 | 10' |
| 12° | 1.022 | 4.810 | 78° |
| 10' | 1.023 | 4.745 | 50' |
| 20' | 1.024 | 4.682 | 40' |
| 30' | 1.024 | 4.620 | 30' |
| 40' | 1.025 | 4.560 | 20' |
| 50' | 1.026 | 4.502 | 10' |
| 13° | 1.026 | 4.445 | 77° |
| 10' | 1.027 | 4.390 | 50' |
| 20' | 1.028 | 4.336 | 40' |
| 30' | 1.028 | 4.284 | 30' |
| 40' | 1.029 | 4.232 | 20' |
| 50' | 1.030 | 4.182 | 10' |
| 14° | 1.031 | 4.134 | 76° |
| 10' | 1.031 | 4.086 | 50' |
| 20' | 1.032 | 4.039 | 40' |
| 30' | 1.033 | 3.994 | 30' |
| 40' | 1.034 | 3.950 | 20' |
| 50' | 1.034 | 3.906 | 10' |
| 15° | 1.035 | 3.864 | 75° |
| | csc θ | sec θ | θ |

| φ | sec φ | csc φ | |
|---|---|---|---|
| 15° | 1.035 | 3.864 | 75° |
| 10' | 1.036 | 3.822 | 50' |
| 20' | 1.037 | 3.782 | 40' |
| 30' | 1.038 | 3.742 | 30' |
| 40' | 1.039 | 3.703 | 20' |
| 50' | 1.039 | 3.665 | 10' |
| 16° | 1.040 | 3.628 | 74° |
| 10' | 1.041 | 3.592 | 50' |
| 20' | 1.042 | 3.556 | 40' |
| 30' | 1.043 | 3.521 | 30' |
| 40' | 1.044 | 3.487 | 20' |
| 50' | 1.045 | 3.453 | 10' |
| 17° | 1.046 | 3.420 | 73° |
| 10' | 1.047 | 3.388 | 50' |
| 20' | 1.048 | 3.356 | 40' |
| 30' | 1.049 | 3.326 | 30' |
| 40' | 1.049 | 3.295 | 20' |
| 50' | 1.050 | 3.265 | 10' |
| 18° | 1.051 | 3.236 | 72° |
| 10' | 1.052 | 3.207 | 50' |
| 20' | 1.053 | 3.179 | 40' |
| 30' | 1.054 | 3.152 | 30' |
| 40' | 1.056 | 3.124 | 20' |
| 50' | 1.057 | 3.098 | 10' |
| 19° | 1.058 | 3.072 | 71° |
| 10' | 1.059 | 3.046 | 50' |
| 20' | 1.060 | 3.021 | 40' |
| 30' | 1.061 | 2.996 | 30' |
| 40' | 1.062 | 2.971 | 20' |
| 50' | 1.063 | 2.947 | 10' |
| 20° | 1.064 | 2.924 | 70° |
| 10' | 1.065 | 2.901 | 50' |
| 20' | 1.066 | 2.878 | 40' |
| 30' | 1.068 | 2.855 | 30' |
| 40' | 1.069 | 2.833 | 20' |
| 50' | 1.070 | 2.812 | 10' |
| 21° | 1.071 | 2.790 | 69° |
| 10' | 1.072 | 2.769 | 50' |
| 20' | 1.074 | 2.749 | 40' |
| 30' | 1.075 | 2.729 | 30' |
| 40' | 1.076 | 2.709 | 20' |
| 50' | 1.077 | 2.689 | 10' |
| 22° | 1.079 | 2.669 | 68° |
| 10' | 1.080 | 2.650 | 50' |
| 20' | 1.081 | 2.632 | 40' |
| 30' | 1.082 | 2.613 | 30' |
| | | | 67° |
| | csc θ | sec θ | θ |

# Natural Secants and Cosecants.

| φ | sec φ | csc φ | |
|---|---|---|---|
| **22°** | | | |
| 30' | 1.082 | 2.613 | 30' |
| 40' | 1.084 | 2.595 | 20' |
| 50' | 1.085 | 2.577 | 10' |
| **23°** | 1.088 | 2.559 | **67°** |
| 10' | 1.088 | 2.542 | 50' |
| 20' | 1.089 | 2.525 | 40' |
| 30' | 1.090 | 2.508 | 30' |
| 40' | 1.092 | 2.491 | 20' |
| 50' | 1.093 | 2.475 | 10' |
| **24°** | 1.095 | 2.459 | **66°** |
| 10' | 1.096 | 2.443 | 50' |
| 20' | 1.097 | 2.427 | 40' |
| 30' | 1.099 | 2.411 | 30' |
| 40' | 1.100 | 2.396 | 20' |
| 50' | 1.102 | 2.381 | 10' |
| **25°** | 1.103 | 2.366 | **65°** |
| 10' | 1.105 | 2.352 | 50' |
| 20' | 1.106 | 2.337 | 40' |
| 30' | 1.108 | 2.323 | 30' |
| 40' | 1.109 | 2.309 | 20' |
| 50' | 1.111 | 2.295 | 10' |
| **26°** | 1.113 | 2.281 | **64°** |
| 10' | 1.114 | 2.268 | 50' |
| 20' | 1.116 | 2.254 | 40' |
| 30' | 1.117 | 2.241 | 30' |
| 40' | 1.119 | 2.228 | 20' |
| 50' | 1.121 | 2.215 | 10' |
| **27°** | 1.122 | 2.203 | **63°** |
| 10' | 1.124 | 2.190 | 50' |
| 20' | 1.126 | 2.178 | 40' |
| 30' | 1.127 | 2.166 | 30' |
| 40' | 1.129 | 2.154 | 20' |
| 50' | 1.131 | 2.142 | 10' |
| **28°** | 1.133 | 2.130 | **62°** |
| 10' | 1.134 | 2.118 | 50' |
| 20' | 1.136 | 2.107 | 40' |
| 30' | 1.138 | 2.096 | 30' |
| 40' | 1.140 | 2.085 | 20' |
| 50' | 1.142 | 2.074 | 10' |
| **29°** | 1.143 | 2.063 | **61°** |
| 10' | 1.145 | 2.052 | 50' |
| 20' | 1.147 | 2.041 | 40' |
| 30' | 1.149 | 2.031 | 30' |
| 40' | 1.151 | 2.020 | 20' |
| 50' | 1.153 | 2.010 | 10' |
| **30°** | 1.155 | 2.000 | **60°** |
| | csc θ | sec θ | θ |

| φ | sec φ | csc φ | |
|---|---|---|---|
| **30°** | 1.155 | 2.000 | **60°** |
| 10' | 1.157 | 1.990 | 50' |
| 20' | 1.159 | 1.980 | 40' |
| 30' | 1.161 | 1.970 | 30' |
| 40' | 1.163 | 1.961 | 20' |
| 50' | 1.165 | 1.951 | 10' |
| **31°** | 1.167 | 1.942 | **59°** |
| 10' | 1.169 | 1.932 | 50' |
| 20' | 1.171 | 1.923 | 40' |
| 30' | 1.173 | 1.914 | 30' |
| 40' | 1.175 | 1.905 | 20' |
| 50' | 1.177 | 1.896 | 10' |
| **32°** | 1.179 | 1.887 | **58°** |
| 10' | 1.181 | 1.878 | 50' |
| 20' | 1.184 | 1.870 | 40' |
| 30' | 1.186 | 1.861 | 30' |
| 40' | 1.188 | 1.853 | 20' |
| 50' | 1.190 | 1.844 | 10' |
| **33°** | 1.192 | 1.836 | **57°** |
| 10' | 1.195 | 1.828 | 50' |
| 20' | 1.197 | 1.820 | 40' |
| 30' | 1.199 | 1.812 | 30' |
| 40' | 1.202 | 1.804 | 20' |
| 50' | 1.204 | 1.796 | 10' |
| **34°** | 1.206 | 1.788 | **56°** |
| 10' | 1.209 | 1.781 | 50' |
| 20' | 1.211 | 1.773 | 40' |
| 30' | 1.213 | 1.766 | 30' |
| 40' | 1.216 | 1.758 | 20' |
| 50' | 1.218 | 1.751 | 10' |
| **35°** | 1.221 | 1.743 | **55°** |
| 10' | 1.223 | 1.736 | 50' |
| 20' | 1.226 | 1.729 | 40' |
| 30' | 1.228 | 1.722 | 30' |
| 40' | 1.231 | 1.715 | 20' |
| 50' | 1.233 | 1.708 | 10' |
| **36°** | 1.236 | 1.701 | **54°** |
| 10' | 1.239 | 1.695 | 50' |
| 20' | 1.241 | 1.688 | 40' |
| 30' | 1.244 | 1.681 | 30' |
| 40' | 1.247 | 1.675 | 20' |
| 50' | 1.249 | 1.668 | 10' |
| **37°** | 1.252 | 1.662 | **53°** |
| 10' | 1.255 | 1.655 | 50' |
| 20' | 1.258 | 1.649 | 40' |
| 30' | 1.260 | 1.643 | 30' |
| | | | **52°** |
| | csc θ | sec θ | θ |

| φ | sec φ | csc φ | |
|---|---|---|---|
| **37°** | | | |
| 30' | 1.260 | 1.643 | 30' |
| 40' | 1.263 | 1.636 | 20' |
| 50' | 1.266 | 1.630 | 10' |
| **38°** | 1.269 | 1.624 | **52°** |
| 10' | 1.272 | 1.618 | 50' |
| 20' | 1.275 | 1.612 | 40' |
| 30' | 1.278 | 1.606 | 30' |
| 40' | 1.281 | 1.601 | 20' |
| 50' | 1.284 | 1.595 | 10' |
| **39°** | 1.287 | 1.589 | **51°** |
| 10' | 1.290 | 1.583 | 50' |
| 20' | 1.293 | 1.578 | 40' |
| 30' | 1.296 | 1.572 | 30' |
| 40' | 1.299 | 1.567 | 20' |
| 50' | 1.302 | 1.561 | 10' |
| **40°** | 1.305 | 1.556 | **50°** |
| 10' | 1.309 | 1.550 | 50' |
| 20' | 1.312 | 1.545 | 40' |
| 30' | 1.315 | 1.540 | 30' |
| 40' | 1.318 | 1.535 | 20' |
| 50' | 1.322 | 1.529 | 10' |
| **41°** | 1.325 | 1.524 | **49°** |
| 10' | 1.328 | 1.519 | 50' |
| 20' | 1.332 | 1.514 | 40' |
| 30' | 1.335 | 1.509 | 30' |
| 40' | 1.339 | 1.504 | 20' |
| 50' | 1.342 | 1.499 | 10' |
| **42°** | 1.346 | 1.494 | **48°** |
| 10' | 1.349 | 1.490 | 50' |
| 20' | 1.353 | 1.485 | 40' |
| 30' | 1.356 | 1.480 | 30' |
| 40' | 1.360 | 1.476 | 20' |
| 50' | 1.364 | 1.471 | 10' |
| **43°** | 1.367 | 1.466 | **47°** |
| 10' | 1.371 | 1.462 | 50' |
| 20' | 1.375 | 1.457 | 40' |
| 30' | 1.379 | 1.453 | 30' |
| 40' | 1.382 | 1.448 | 20' |
| 50' | 1.386 | 1.444 | 10' |
| **44°** | 1.390 | 1.440 | **46°** |
| 10' | 1.394 | 1.435 | 50' |
| 20' | 1.398 | 1.431 | 40' |
| 30' | 1.402 | 1.427 | 30' |
| 40' | 1.406 | 1.423 | 20' |
| 50' | 1.410 | 1.418 | 10' |
| **45°** | 1.414 | 1.414 | **45°** |
| | csc θ | sec θ | θ |

# EXPLANATION OF THE TABLES.

## § 1. TABLES IN GENERAL.

*a*. One quantity is said to be a **function** of another, when the former quantity is regarded as determined by the latter, according to some rule or formula. E. g. $x^2$, $\sqrt{x}$, log $x$, sin $x$, log sin $x$, are all called functions of $x$. A **mathematical table** is an orderly arrangement of the values of some function for certain selected values of the quantity by which it is regarded as determined. The successive values of the latter quantity are assumed arbitrarily, and generally at equal intervals ; and this quantity is called the **argument** of the table. Some functions require several independent quantities for their determination ; and the corresponding tables are tables of several arguments. Thus, a multiplication-table is a table of *two* arguments ; namely, the two factors.

*b*. A table may be used in two ways : **directly** and **inversely**. The direct use of the table consists in finding the value of the function for an assumed value of the argument ; the inverse use, in finding the value of the argument for an assumed value of the function.

*c*. Before beginning to use any table, the student should give attentive consideration to its arrangement, and to the best mode of employing it with accuracy and ease. Every feature of it should be carefully examined, and the explanations which are attached to it should be fully mastered. The time thus spent will be time gained, contributing not only to power in computation, but also, very materially, to the thorough practical knowledge of the nature of the tabulated functions.

## § 2. INTERPOLATION.

*a*. **Interpolation** consists in finding the value of one of the two quantities, argument and function, for an assumed value of the other quantity, lying between two successive tabulated values. Most mathematical tables are so constructed as to admit of interpolation by the principle that *corresponding non-tabulated values of the function and argument lie between corresponding tabulated values and divide the differences between them in the same ratio.* This is the principle of **proportional parts**. Let $x_1$ and $x_2$ be two successive tabulated values of the argument of a table, and $u_1$ and $u_2$ the correspond-

29

ing values of the function. Then, $x_2 - x_1$ and $u_2 - u_1$ are called corresponding tabular differences. We shall denote these differences by $\Delta x$ and $\Delta u$. If, now, $x$ and $u$ are corresponding values of the function and argument, of which one is known to lie between the two above-cited tabulated values of the same quantity, the principle of proportional parts is that if

$$\lambda = \frac{x - x_1}{\Delta x}, \qquad\qquad \lambda' = \frac{x_2 - x}{\Delta x} = 1 - \lambda.$$

$$\mu = \frac{u - u_1}{\Delta u}, \qquad\qquad \mu' = \frac{u_2 - u}{\Delta u} = 1 - \mu,$$

then (to the limit of accuracy belonging to the table)

$$\lambda = \mu, \qquad\qquad \lambda' = \mu',$$

or,

$$u = u_1 + \lambda \, \Delta u = u_2 - \lambda' \, \Delta u,$$
$$x = x_1 + \mu \, \Delta x = x_2 - \mu' \, \Delta x.$$

Thus, the required value of the function or argument may be obtained by applying a **correction** to either of the two tabulated values between which the required value lies. In computing this correction, the *signs* of the differences employed must be carefully observed. If $x_1$ and $x_2$ are so chosen as to make $\Delta x$ positive, $\Delta u$ may be either positive or negative. In the former case, the function is said to be *increasing ;* in the latter, *decreasing.*

*b.* Either of the two formulas given above for finding $u$ may be employed, in interpolation, in the direct use of the table; either of the formulas for $x$ may be employed in the inverse use of the table. In most tables, $\Delta x =$ *one unit* in the last numeral place of the tabulated values of $x$. Hence $\lambda$ is composed of *the figures which follow that numeral place* in the given non-tabulated value of the argument, preceded by a decimal-point; while $\lambda'$ is the *complement* of $\lambda$ (that is, can be found by subtracting from 9 each figure of $\lambda$ except the last, and subtracting that from 10). The correction for $u$ is, therefore, found simply by *multiplying* the figures in question into $\Delta u$, and pointing off according to the case; $x$ will be corrected by *annexing* to $x_1$ the figures of $\frac{u - u_1}{\Delta u}$, or the figures complementary to $\frac{u_2 - u}{\Delta u}$.

*c.* In some of the tables of this collection will be found, set against each value of the function, *a number in small type*, which shows what $\Delta u$ would be if the function varied through a whole interval corresponding to $\Delta x$ at the same rate at which it is changing when it passes through the value against which this number is set. This number may be called the *rate of difference,* or simply the *difference,* of $u$, and may be substituted for $\Delta u$ in the formulas of interpolation. But, in that case, we ought to work from the NEAREST tabulated value of $x$ or $u$; that is, from $x_1$ or $u_1$ when $\lambda$ or $\mu < 0.5$, and from $x_2$ or $u_2$ when $\lambda'$ or $\mu' < 0.5$. (See examples in the explanation of the table of *Logarithms of Circular Functions.*)

*d. An interpolated value of the function should not be carried out beyond the last numeral place of the tabulated value from which it is computed ;* so that, in finding $\lambda \Delta u$ or $\lambda' \Delta u$, we should reject the decimal part of the product, $\Delta u$ being regarded as an integer. Owing to the combination of the figures rejected in the correction and those omitted in the tabulated value of the function, an interpolated value is liable to an error of $\pm 1$ in the last figure.

*The number of figures annexed to the tabulated value of the argument, in inverse interpolation, should be less by one than the number of figures contained in $\Delta u$.* It is sometimes, indeed, made *equal* to the latter number (and will always be, if $\Delta u$ consists of only one figure); but, in that case, the *last figure* must be regarded as *uncertain.* When the given value of the function is the result of computation, of course this uncertainty may extend back to earlier figures.

*e.* In taking the correction of either the function or the argument only to a certain number of figures, we must observe the following rule, which is a universal rule of computation: —

**Whenever figures are neglected at the end of a number, if the figures neglected amount to more than half a unit in the place of the last figure retained, the last figure retained must be increased by 1.** E. g. $27.528 = 27.53$ to the nearest hundredth $= 27.5$ to the nearest tenth $= 28$ to the nearest unit $= 30$ to the nearest ten.

*f.* The various rules of interpolation will be found to be fully exemplified below, in the explanations of the tables of *Logarithms* and *Logarithms of Circular Functions.*

*g.* In interpolating in some tables (e. g. in VLACQ's great ten-place table), we must have regard to **second differences**, or differences between differences. In this case, we add to the above formulas for $u$ the term

$$- \tfrac{1}{2} \lambda \lambda' \Delta^2 u,$$

where $\Delta^2 u$ denotes the second difference of $u$, taken positively when $\Delta u$ is increasing. The greatest value of this term is *one eighth* of $\Delta^2 u$, so that it is insignificant when $\Delta^2 u < 4$. *In the present tables this term may always be neglected;* although it is useful as measuring the extent of error, and may occasionally guide the judgment of the computer when the fractional part of the correction is equal, or nearly equal, to 0.5. But where such nicety of work seems to be called for, it is best to use a table of a larger number of places.

## § 3. PROPORTIONAL PARTS.

*a.* The table of **Proportional Parts** (folded page) may be used in connexion with any other table, as an aid in *interpolation.* It contains the product of every integer from 1 to 100 by every *tenth* from 0.1 to 0.9. If the multiplier consists of one figure in any other numeral place, it is only necessary to change the position of the decimal-point in the product. To multiply a number of two figures by any decimal whatever, we must find the products which correspond to the successive figures of the multiplier, and add them together. The decimal part of the result is generally to be discarded, and in that case the general rule given above (in § 2, *e*) must be observed. Thus, let it be required to find $0.619 \times 37$. Looking in the column belonging to 37, we find

$$0.6 \quad \times 37 = 22.2$$
$$0.01 \quad \times 37 = 0.37$$
$$0.009 \times 37 = \underline{0.333}$$
$$\therefore \quad 0.619 \times 37 = 23.$$

In like manner, we find

| | | |
|---|---|---|
| $0.27 \times 15 = 4,$ | $0.59 \times 73 = 43,$ | $0.78 \times 69 = 54,$ |
| $0.96 \times 84 = 81,$ | $0.36 \times 57 = 21,$ | $0.289 \times 51 = 15,$ |
| $0.483 \times 93 = 45,$ | $0.374 \times 82 = 31,$ | $0.053 \times 68 = 4.$ |

*b.* This table can also be used *inversely*. Thus, let it be required to find, to two decimal-places, what part 36 is of 79. Looking in the column of 79, we find

$$0.4 \times 79 = \frac{36}{\underset{\overline{\phantom{00}4.4}}{31.6}}$$

$$0.06 \times 79 = 4.74 \text{ (the } \textit{nearest} \text{ product)}$$

$$\therefore \frac{36}{79} = 0.46.$$

In like manner, we find

$$\frac{29}{68} = 0.43, \quad \frac{72}{89} = 0.81, \quad \frac{31}{98} = 0.32, \quad \frac{26}{71} = 0.37, \quad \frac{45}{57} = 0.79, \quad \frac{11}{37} = 0.30.$$

A little practice will enable the student to use this table easily and rapidly.

## § 4. LOGARITHMS.

*a.* **Denary**, or **Briggsian**, logarithms, being those employed in actual computation, are always referred to, in this collection of tables, when the term *logarithm* is used without qualification. The **characteristic**, or integral part, of the denary logarithm of a number depends only on the position of the first significant figure of the number relatively to the units' place, and may be found by a well-known rule; the **mantissa**, or fractional part, depends only on the series of significant figures which compose the number, and is the only part of the logarithm for which it is necessary to employ a table. A table of logarithms is complete, to an assigned number of places, if it gives (explicitly or by interpolation) to that number of places the mantissa of the logarithm of every possible series of significant figures. Denary logarithms are, in general, incommensurable numbers, and cannot, therefore, be exactly expressed in figures. They are variously given, in different tables, to ten, seven, six, five, four, and three places of decimals. Four-place logarithms are sufficient for the ordinary purposes of engineering, navigation, the work of the physical and chemical laboratory, and many of the subordinate computations of astronomy; and, in most of these cases, are all that the accuracy of the data will justify us in using. Seven places are, however, needful for the more accurate kinds of astronomical and geodetic work.

*b.* If one number is the logarithm of another, the second number is called the **antilogarithm** of the first. This relation is denoted by the symbol $\log^{-1}$. Thus, if $u = \log x$, then $x = \log^{-1} u$. In an ordinary table of logarithms, the *argument* is the antilogarithm, which is tabulated to a greater or less number of figures, according to the number of places to which the logarithm is given, and the *function* is the mantissa of the loga-rithm, which we often speak of simply as the *logarithm*.

### To find the logarithm of any number.

*c.* If the number consists of *three significant figures*, seek the first two significant figures in the first column of the table of **Logarithms** (pp. 2, 3), and the third at the top of the table. In the line and column thus de-termined will be found the mantissa of the required logarithm, printed without the decimal-point. Find the characteristic by the rule, and prefix it, with the decimal-point, to the mantissa. E. g., log 2870 = 3.4579. If the given number has *less than three* significant figures, fill it out to three figures by anexing a zero or zeros. E. g., log 0.35 = log 0.350 = 9.5441 — 10, log 6 = log 6.00 = 0.7782. If the number has *more than three* significant figures, its logarithm must be found by one of the formulas of interpolation given above. The rule is : — *Find the logarithm of the first three significant figures of the given number and also that of the next following number of three figures* (1000 following 999) ; *then apply to* EITHER *of these two logarithms a correction, obtained by multiplying the difference between them by the difference between the given number and the three-figure number which corresponds to the loga-rithm chosen to be corrected, and rejecting (with due attention to the rule of § 2, e) as many figures at the end of the product as are contained in the latter difference.* The table of **Proportional Parts** may be employed in performing the multiplications. Thus, to find log 5668.4. Using the notation of the formu-las of interpolation, and remembering that the place of the decimal-point in the given number may be disregarded in finding the mantissa of the required logarithm, we have

$$x_1 = 566, \qquad u_1 = \text{mant log } x_1 = 7528,$$
$$x_2 = 567, \qquad u_2 = \text{mant log } x_2 = 7536,$$
$$\Delta x = \quad 1, \qquad \Delta u = \qquad\qquad 8 ;$$

so that log $x$ may be found by either of the following methods : —

$$\lambda = 0.84, \qquad \lambda \Delta u = 6.72 = 7 \text{ to units}, \qquad u = 7528 + 7 = 7535;$$

or,

$$\lambda' = 1 - \lambda = 0.16, \qquad \lambda'\Delta u = 1.28 = 1 \text{ to units}, \qquad u = 7536 - 1 = 7535.$$
$$\therefore \text{ log } 5668.4 = 3.7535.$$

Let the beginner find the following logarithms by this method : —

| | | | |
|---|---|---|---|
| log 59.43 | = 1.7740, | log 0.0081472 | = 7.9110 — 10, |
| log 284.8 | = 2.4545, | log 572820 | = 5.7581, |
| log 0.073748 | = 8.8678 — 10, | log 0.50167 | = 9.7004 — 10, |
| log 3.1607 | = 0.4998, | log 99968. | = 4.9999. |

*The interpolated logarithm should never be carried to more than four decimal-places.*

*d.* The work of interpolation may be shortened by using the column of proportional parts, marked P. P., on the right of the table. In using this column, one must work from the three-figure number NEAREST to the given

## Explanation of the Tables.

number (in the above example, from 567, not from 566). If the given number has only four figures, so that λ or λ' has only one figure, then the correction will be found in the column P. P., under λ or λ' (according as we are working from the number below or the number above the given number), and in the same line with the logarithm to be corrected. If the given number has more than four figures, the correction must be estimated by the observation of the corrections which correspond to the figures below and above the first figure of λ or λ'. E. g.

mant log 2848 = mant log 285 — cor. for .2 = 4548 — 3 = 4545;
mant log 56684 = mant log 567 — cor. for .16 = 7536 — 2 = 7534

In the last case the correction is either 1 or 2, and, since .16 is nearer .20 than .10, we choose the correction belonging to .20. Larger tables show that the mantissa of the required logarithm, to five places, is 75346; so that the value found by the column P. P. is here nearly as accurate as that obtained by computation. There is a slightly greater liability to error when we use the column P. P. than when we interpolate by computation; but the disadvantage is generally insignificant. The last figure of an interpolated logarithm obtained from any table may always be one unit in error. E. g.: the true mant log of 57282 to five places is 75802; and this is a case in which the column P. P. gives a better result than computation.

The student is advised now to find all the logarithms in the above list by using the column P. P.

*e. If the first figure of the given number is* 1, it will be found tabulated to *four* figures in pp. 4, 5. The correction for a fifth and following figures may be found by the method of interpolation explained in *c.* As the differences are always small on these pages, and the corrections easily computed, the column P. P. is not here given ; but, to facilitate taking the last difference, we have printed at the end of each line, under the heading 10, the first logarithm of the following line. Let the student find the following logarithms : —

log 11.737  = 1.0696,        log 0.00100066 = 7.0003 — 10,
log 0.15703 = 9.1960 — 10,    log 18597.    = 4.2694.

## To find the antilogarithm of any logarithm.

*f.* It is enough to explain the way of finding the series of significant figures which compose the antilogarithm, by means of the mantissa of the given logarithm ; the pointing off of the antilogarithm being determined, according to rule, by the given characteristic. If the mantissa of the given logarithm is contained in the table, the required antilogarithm is at once found by inspection. Otherwise, we must resort to the formulas of interpolation, which give the following rule : — *Find two successive tabulated logarithms* ($u_1$ *and* $u_2$) *between which the given logarithm* ($u$) *lies ; then divide the difference between either of these tabulated logarithms and the given logarithm* ($u — u_1$ *or* $u_2 — u$) *by the difference between the tabulated logarithms* ($\Delta u$), *carry out the quotient to the* NEAREST *tenth* (that is, to one figure, which may be 0), *and add it to or subtract it from the antilogarithm* ($x_1$ *or* $x_2$) *of the tabulated logarithm* ($u_1$ *or* $u_2$) *with which the given logarithm has been compared.* The antilogarithm is always a figure *annexed* to the three or four tabulated figures of $x_1$.

The division should not generally be carried beyond one figure. Even the first figure is, in most cases, somewhat uncertain. If the mantissa of the given logarithm is less than 3010, pp. 4, 5 should be used. On pp. 2, 3, the column P. P. may be employed. Let it be required to find $\log^{-1} 1.5284$. We find

$$u_1 = 5276, \qquad x_1 = 337, \qquad u - u_1 = 8, \qquad u_2 - u = 5,$$

$$u_2 = 5289, \qquad x_2 = 338, \qquad \mu = \mu \,\Delta x = \frac{8}{13} = 0.6..,$$

$$\Delta u = \quad 13, \qquad \Delta x = \quad 1, \qquad \mu' = \mu' \Delta x = \frac{5}{13} = 0.4..,$$

$$x = 337 + 0.6 = 338 - 0.4 = 337.6;$$
$$\therefore \ \log^{-1} 1.5284 = 33.76.$$

More briefly, looking along the line of $u_2 = 5289$ for $5 = u_2 - u$ in column P. P., we find that 5 corresponds to the correction 4, which gives at once the required number. In like manner, the student may find

| | |
|---|---|
| $\log^{-1} 1.9155 = 82.32,$ | $\log^{-1} (5.8760 - 10) = 0.00007517,$ |
| $\log^{-1} 3.8291 = 6747,$ | $\log^{-1} (9.5727 - 10) = 0.3738,$ |
| $\log^{-1} 0.1548 = 1.4283,$ | $\log^{-1} (8.2731 - 10) = 0.018755.$ |

*g.* The convenient usage of *making negative characteristics positive*, by the addition of 10, is followed, throughout the present collection of tables, whenever logarithms are printed with their characteristics. This must be always understood, though no explicit reference be made to it in the explanation of the table.

## § 5. LOGARITHMS OF SUMS AND DIFFERENCES.

*a.* This is one form of a table devised by GAUSS to facilitate finding the logarithm of the *sum* or *difference* of two numbers which are themselves given only by their logarithms. The *argument* of the table is any logarithm, and may be called log $x$; the *function* tabulated is then log $(x + 1)$. It follows that, if the function is denoted by log $x$, the argument is log $(x - 1)$. The function may be called the **Gaussian** of the argument, and the argument the **anti-Gaussian** of the function; and the symbols ⑤ and ⑤$^{-1}$ may be used to denote these relations. Thus we have

$$\log (x + 1) = \textcircled{5} \log x, \qquad \log (x - 1) = \textcircled{5}^{-1} \log x.$$

*b.* **To find the Gaussian of a given logarithm.** Seek the characteristic of the given logarithm (increased by 10 if negative) at the top of the table, and the first two figures of the mantissa in the left-hand column. If the third and fourth figures of the mantissa are zero, the Gaussian will be found in the column and line thus determined; otherwise, it can be obtained by the method of interpolation which has been fully explained in § 2. In three columns of the table, the *rate of difference* of the Gaussian is printed in small type after the value of the function, and may be used instead of the tabular difference of the Gaussian *through half the tabular interval before and after* the value to which it is attached, as explained in

§ 2, c, and completely illustrated below, in the explanation of the table of *Logarithms of Circular Functions*. The table of Proportional Parts may be employed in computing the corrections. Examples:—

$$\text{⑤ } 1.0900 = 1.1295, \qquad \text{⑤ } (7.5265 - 10) = 0.0015,$$
$$\text{⑤ } 3.8129 = 3.8130, \qquad \text{⑤ } (9.6431 - 10) = 0.1582.$$

If $\log x < 0.0000 - 10$,    $\log (x + 1) = 0.0000$ to four places;
if $\log x > 4.0000$,    $\log (x + 1) = \log x$ to four places.

c. **To find the anti-Gaussian of a given logarithm.** Seek, in the body of the table, two successive logarithms between which the given logarithm lies, and then find the corresponding value of the argument by interpolation. Examples:—

$$\text{⑤}^{-1} 1.0900 = 1.0597, \qquad \text{⑤}^{-1} 0.1051 = 9.4373 - 10,$$
$$\text{⑤}^{-1} 3.8129 = 3.8128, \qquad \text{⑤}^{-1} 1.0216 = 0.9782.$$

d. **To find the logarithm of the sum or difference of the anti-logarithms of two given logarithms.** If $m$ and $n$ are two numbers,

$$m + n = n\left(\frac{m}{n} + 1\right), \qquad m - n = n\left(\frac{m}{n} - 1\right),$$

$$\log (m + n) = \log n + \log\left(\frac{m}{n} + 1\right) = \log n + \text{⑤} \log \frac{m}{n},$$

$$\log (m - n) = \log n + \log\left(\frac{m}{n} - 1\right) = \log n + \text{⑤}^{-1} \log \frac{m}{n}.$$

Example:—

Given $a = 4.142$, $b = 2.399$; to find $\sqrt{(a^2 + b^2)}$ and $\sqrt{(a^2 - b^2)}$.

| | | | |
|---|---|---|---|
| $\log a$ | $= 0.6172,$ | $\log b$ | $= 0.3800,$ |
| $\log a^2$ | $= 1.2344,$ | $\log b^2$ | $= 0.7600,$ |
| $\log \dfrac{a^2}{b^2}$ | $= 0.4744;$ | | |
| $\text{⑤} \log \dfrac{a^2}{b^2}$ | $= 0.6000,$ | $\text{⑤}^{-1} \log \dfrac{a^2}{b^2}$ | $= 0.2970,$ |
| $\log b^2$ | $= 0.7600,$ | $\log b^2$ | $= 0.7600,$ |
| $\log (a^2 + b^2)$ | $= 1.3600,$ | $\log (a^2 - b^2)$ | $= 1.0570,$ |
| $\log \sqrt{(a^2 + b^2)}$ | $= 0.6800,$ | $\log \sqrt{(a^2 - b^2)}$ | $= 0.5285,$ |
| $\sqrt{(a^2 + b^2)}$ | $= 4.787;$ | $\sqrt{(a^2 - b^2)}$ | $= 3.377.$ |

## § 6. CIRCULAR, OR TRIGONOMETRIC, FUNCTIONS: NATURAL VALUES.

a. Three tables of the natural values of the trigonometric functions are given on pp. 22-27. Each table is broken up into six divisions, and occupies two pages. The argument is the angle, which is tabulated at intervals of 10′ from 0° to 90°. Angles in the *first half of the quadrant* will be found in the *left-hand column* of the several divisions of the table, and for those angles the names of the functions are to be taken from the *top* of the page; angles in the *second half of the quadrant* are to be found in the *right-hand column* of the table, and for those angles the names of the functions are to be taken from the *bottom* of the page. The *angles* standing at the right and left in the same line are *complements* of each other; and the names of the *functions* at the top and bottom of the same column are *complementary.* The value of any of the functions for a non-tabulated angle, or the value of the angle for a non-tabulated value of one of the functions, can be found by the method of interpolation explained in § 2. The precepts of § 2, *d, e*, should be observed in computing the corrections. The tabulated values of the functions are generally given to four significant figures; but, in the tables of tangents and secants, they are sometimes given to a less number of figures (to avoid errors in interpolation), and are sometimes omitted altogether. In these cases the functions can be best found by finding their logarithms by the table of *Logarithms of Circular Functions* (see § 7), and then the numbers corresponding by the table of *Logarithms.*

*b.* To find any function of an angle *greater than* 90°, we must subtract from the given angle the greatest multiple of 90° which it contains; if an *even* multiple has been subtracted, we look out the required function of the remainder; if an *odd* multiple, the complementary function; and we then fix the *sign* of the function by considering the *quadrant* in which the given angle lies. For a *negative* angle, we find the required function of the corresponding positive angle, and then fix its sign by considering the quadrant of the angle.

*c.* Examples of the use of these tables : —

sin 77° 37′  =  0.9767,  tan  53° 04′ =  1.330,  sec  68° 45′ =  2.759,
cos 16° 19′  =  0.9597,  ctn  3° 18′ =  17.4,  csc  55° 13′ =  1.217 ;

sin 257° 37′  = —0.9767,  tan  93° 18′ = —17.4,  sec 325° 13′ =  1.217,
cos 163° 41′  = —0.9597,  ctn 323° 04′ = —1.330,  csc 158° 45′ =  2.759 ;

sin (—257° 37′) =  0.9767, tan (— 93° 18′) = 17.4,  sec (—325° 13′) =  1.217,
cos (—163° 41′) = —0.9597, ctn (—323° 04′) = 1.330, csc (—158° 45′) = —2.759 ;

sin⁻¹ 0.2000    =   11° 32′ or =   168° 28′ or =   371° 32′, etc.,
cos⁻¹ (—0.3542) =  110° 45′ or =   249° 15′ or =   830° 45′, etc.,
tan⁻¹ (—4.570)  =  102° 21′ or =   282° 21′ or =   —77° 39′, etc.,
ctn⁻¹ 0.3163    =   72° 27′ or =   252° 27′ or =  —107° 33′, etc.,
sec⁻¹ 5.000     =   78° 28′ or = —78° 28′ or = ±281° 32′, etc.,
csc⁻¹ (—3.529)  = —16° 28′ or =   196° 28′ or = —163° 32′, etc.

## § 7.  LOGARITHMS OF CIRCULAR FUNCTIONS.

**To find the logarithm of any circular function of a given angle.**

*a.* If the angle is *less than* 6°, the part of the table which occupies the upper half of p. 10 may be used. (See also *g*.) The *left-hand division* of this part of the table gives the values of a logarithm S (the characteristic and the first two figures of the mantissa being printed at the head of the column), with the angular limits between which each value may be used. Thus, for all positive angles less than 1° 51′.479, S = 6.4637 ; for all angles between 1° 51′.479 and 2° 49′.567, S = 6.4636; etc. The *next following division* gives, in like manner, the values of a logarithm T. We must find *the logarithm of the angle*, reduced to minutes and decimals of a minute, and must then apply the formulas : —

$$\log \sin \phi = \log (\phi \text{ in minutes}) + S - 10,$$
$$\log \tan \phi = \log (\phi \text{ in minutes}) + T - 10.$$

*The two right-hand divisions* of this part of the table give the values of the log sec, with the angular limits for each value. The logarithms of the cosine, cotangent, and cosecant are the *arithmetical complements* (— 10) of the logarithms of the secant, tangent, and sine, respectively. Example : —

l sin 3° 15′.23 = 8.7541,  l tan 3° 15′.23 = 8.7548,  l sec 3° 15′.23 = 0.0007,
l csc 3° 15′.23 = 1.2459,  l ctn 3° 15′.23 = 1.2452,  l cos 3° 15′.23 = 9.9993 ;

the negative characteristics being here, as in the following examples, made positive by the addition of 10.

*b.* If the angle is *acute and greater than* 84°, we must take its *complement*, and then seek the *function complementary to that required*, for the angle thus obtained, by the method just expounded. Example : —

l sin 86° 44′.77 = 9.9993, l tan 86° 44′.77 = 1.2452, l sec 86° 44′.77 = 1.2459,
l csc 86° 44′.77 = 0.0007, l ctn 86° 44′.77 = 8.7548, l cos 86° 44′.77 = 8.7541.

*c.* If the angle is contained *between 6° and 84°*, we use the main part of the table, occupying the lower half of p. 10 and pp. 11–15. The angle is tabulated at intervals of 10′, from 0° to 45° in the left-hand column of the table, and from 45° to 84° in the right-hand column. The names of the functions are to be taken from the *tops* of the columns, when the angle is on the *left ;* and from the *bottoms* of the columns, when the angle is on the *right*. The angles on the right and left of any line and the names at the top and bottom of any column have the same relation to each other as in the tables of *Natural Values* (§ 6). The true characteristic in the first, third, and sixth columns is —1, but is printed 9. The six columns are arranged in pairs. The two functions in each pair of columns are *reciprocal* to each other ; and the logarithms are therefore *complementary*, and their differences are equal in value, with opposite signs. Down the middle of each double column are printed, in small type, the *rates of difference* of the logarithms in that double column. Each value of this rate may be used in interpolation, instead of Δ *u, through half the interval before and after the line on which it stands*, as stated in § 2, *c.* Thus, in finding the logarithms of the circular functions of any angle between 25° 25′ and 25° 35′ we work from the values corresponding to 25° 30′, the *nearest* tabulated angle ; and compute the corrections by taking proportional parts of 26, 33, and 6, for the three pairs of functions. In applying the corrections, we must carefully observe, for each function, *whether the function is increasing or decreasing.*

For example, let the logarithms of the circular functions of 25° 27′.4 = 25° 30′ — 02′.6 be required. We find

| | | |
|---|---|---|
| l sin 25° 30′  = 9.6340 | l tan 25° 30′  = 9.6785 | l sec 25° 30′  = 0.0445 |
| 0.26 × 26 =  7 | 0.26 × 33 =  9 | 0.26 × 6 =  2 |
| l sin 25° 27′.4 = 9.6333, | l tan 25° 27′.4 = 9.6776, | l sec 25° 27′.4 = 0.0443, |

# Logarithms of Circular Functions.

l csc 25° 30′ = 0.3660    l ctn 25° 30′ = 0.3215    l cos 25° 30′ = 9.9555
             7                 9                 2

l csc 25° 27′.4 = 0.3667,    l ctn 25° 27′.4 = 0.3224,    l cos 25° 27′.4 = 9.9557.

In like manner, we have

l sin 74° 46′ = 9.9845,    l tan 74° 46′ = 0 5649,    l sec 74° 46′ = 0.5804,
l csc 74° 46′ = 0.0155,    l ctn 74° 46′ = 9.4351,    l cos 74° 46′ = 9.4196.

*d.* If the angle is *greater than* 90°, or *negative*, we must use the method explained in § 6, *b*, for the tables of *Natural Values* of the circular functions. When the natural value of a circular function is negative, this should be indicated by writing the letter *n* after its logarithm. Examples:

l sin 105° 14′ = 9.9845,           l tan 105° 14′ = 0 5649 *n*,
        l sec 105° 14′ = 0.5804 *n*,

l csc 164° 46′ = 0.5804,           l ctn 164° 46′ = 0.5649 *n*,
        l cos 164° 46′ = 9.9845 *n*,

l sin (—74° 46′) = 9.9845 *n*,      l tan (—105° 14′) = 0.5649,
        l cos (—394° 46′) = 9.4196.

**Given the logarithm of any circular function, to find the value of the corresponding angle.**

*e.* If the given logarithm lies *without the limits of the main part of the table*, the upper part of p. 10 may be used. If the given logarithm is a *log sin less than* 9.0192, or a *log tan less than* 9.0216, subtract from it the proper value of S or T (or add the arithmetical complement), and the remainder is the log of the required angle in minutes. The limiting values of the log sin and log tan for each value of S and T are given in the table. If the given log is a *log csc greater than* 0.9808, or a *log ctn greater than* 0.9784, its arithmetical complement will be a log sin less than 9.0192, or a log tan less than 9.0216. If the given log is a *log sec less than* 0.0024, the limits between which the required angle lies are given by the table; the angle may have any value between these limits, and is not therefore very closely determined. If the given log is a *log cos greater than* 9.9976, its arithmetical complement is a log sec less than 0.0024.

If the given log is a *log sin, log tan, or log sec greater than* 9.9976, 0.9784, *or* 0.9808 (respectively), or a *log csc, log ctn, or log cos less than* 0.0024, 9.0216, *or* 9.0192 (respectively), we must change the name of the function to the *complementary name* (sin to cos, etc.), *find the corresponding angle* as above, and take the *complement* of the angle thus found. Examples: —

(log sin)⁻¹ 8.9542 = 5° 09′.8,     (log ctn)⁻¹ 2.0531 = 0° 30′.42,
(log ctn)⁻¹ 9.0024 = 84° 15′.5,    (log sin)⁻¹ 9.9983 = 84° 56′±4¼′.

*f.* If the given logarithm is contained *within the limits of the main part of the table*, the required angle is found by ordinary interpolation; and we may use the printed rate of difference as the value of $\Delta u$, working in each case from the *nearest* tabulated value. The angle should be found to the *nearest minute*, or, when the difference exceeds 100, to the nearest *tenth* of a minute. But in the right-hand pair of columns, the last figure of the angle thus found will generally be uncertain. Examples: —

Let it be required to find (log sec)$^{-1}$ 0.0343; *i. e.* the angle of which the log sec is 0.0643. The nearest tabulated log sec is 0.0647. We have, then,

$$(\log \sec)^{-1} \, 0.0647 = 30° \, 30', \ u_2 - u = 4, \ \Delta u = 7, \ \frac{4}{7} = 0.6,$$

$$\therefore (\log \sec)^{-1} \, 0.0643 = 30° \, 30' - 06' = 30° \, 24'.$$

In like manner, let the student find

$(\log \sin)^{-1} \, 9.5663 = 21° \, 37',$       $(\log \operatorname{ctn})^{-1} \, 0.0496 = 41° \, 44',$

$(\log \cos)^{-1} \, 9.9188 = 33° \, 58'$ or $57',$     $(\log \sec)^{-1} \, 0.2272 = 53° \, 39',$

$(\log \tan)^{-1} \, 0.7507 = 79° \, 56',$       $(\log \csc)^{-1} \, 0.1433 = 45° \, 58'.$

The angle may also be found by the next following table.

*g.* Pp. 8 and 9 may also be used for angles less than 6° or greater than 84°.

E.g. l sin 4° 03′.4 = 8.8497, l tan 4° 03′.4 = 8.8508, l sec 4° 03′.4 = 0.0011, l csc 4° 03′.4 = 1.1503, l ctn 4° 03′.4 = 1.1492, l cos 4° 03′.4 = 9.9989.

## § 8. INVERSE CIRCULAR FUNCTIONS.

*a.* The table having this heading (pp. 16–18) is a table for finding the angle which corresponds to the given logarithm of a circular function. The logarithm (increased by 10) is the *argument* of the table, and is to be regarded as given to *four places* of decimals. It is tabulated at intervals of 0.0100 from 9.0000 to 0.0000 through the first page of the table, then at intervals of 0.0010, and in the last two divisions at intervals of 0.0001. The characteristic of the argument is printed at the head of the column. The figures supposed to follow the printed figures in the values of the argument are *zeros*. Thus, the first value is 9.0000, the next 9.0100, etc. The angle is given, for convenience of interpolation, in degrees and decimals of a degree. When found, it is easily reduced to degrees and minutes, if that is necessary, and should, in general, be taken *only to the nearest minute*. The angle under the heading sin$^{-1}$ *u* is that angle of which the corresponding value of the argument, log *u*, is the log sin; etc. In interpolating in this table, we may use the *printed rate of difference* instead of $\Delta u$, working from the *nearest* tabulated value of the argument, and carefully observing whether the tabulated angle ought to be *increased* or *diminished*. *When the printed rate of difference is omitted*, this is because the interval is too great to admit of accurate interpolation. In this case, we must resort to those later divisions of the table in which the argument is tabulated at smaller intervals. *When the last figure of the tabulated angle is printed in small type*, this shows that that figure is uncertain, if the logarithm is given to only four places; that is, that there is a possible variation, on each side of the tabulated angle, as great as *half a unit* in the place of the figure so printed. For example, if log *u* = 9.9000, we find the last figures of sin$^{-1}$ *u* and cos$^{-1}$ *u* to be printed in small type. Now, seven-place tables show that (log sin)$^{-1}$ 9.8999500 = 52°.581, while (log sin)$^{-1}$ 9.9000500 = 52°.600. But 9.9000 may represent any logarithm between these; and hence the corresponding angle, in this case, admits a like variation, while cos$^{-1}$ *u* may have any value between 37°.419 and 37°.400.

Neither of these difficulties presents itself in finding an angle from its log tan or log ctn. If log $u = 9.9000$, $\tan^{-1} u$ can only vary from $38°.458$ to $38°.464$.

The angle found by interpolation should be carried out *only to the nearest hundredth of a degree, in any case.* The last column of the table shows that the angle is not always determined even to the nearest *tenth.*

*b.* *If the characteristic of the given logarithm is* 0, we must take its *arithmetical complement,* which will be the logarithm of the *reciprocal* function of the same angle. The angle can then be found by the table.

*c.* If the given logarithm is less than 9.0000, or greater than 0.0000, the tables in the upper part of p. 10 may be used, as explained in § 7, *e*; or pp. 8, 9.

*d.* Let us find by this table the angles sought above, in § 7, *f.* We have, in the case of the first example,

$$(\log \sec)^{-1}\, 0.0643 = (\log \cos)^{-1}\, 9.9357.$$

Then the table gives

$$(\log \cos)^{-1}\, 9.9360 = 30°.35, \qquad \Delta u = 23,$$
$$0.3 \times 0.23 \;=\; .07$$
$$\therefore\; (\log \cos)^{-1}\, 9.9357 = 30°.42 = 30°\, 25'.$$

In fact, the limits of the angle are $30°\, 24'.2$ and $30°\, 25'.6$, the mean value being $30°\, 24'.9$. In this case, the present table gives a better value than the other; but both values are admissible.

In like manner, we have

$$(\log \sin)^{-1}\, 9.5663 = 21°.81 - 20 = 21°.61 = 21°\, 37',$$
$$(\log \cos)^{-1}\, 9.9188 = 33°.92 + .04 = 33°.96 = 33°\, 58',$$
$$(\log \tan)^{-1}\, 0.7507 = 79°.92 + .02 = 79°.94 = 79°\, 56',$$
$$(\log \ctn)^{-1}\, 0.0496 = 41°.71 + .03 = 41°.74 = 41°\, 44',$$
$$(\log \sec)^{-1}\, 0.2272 = 53°.64 + .02 = 53°.66 = 53°\, 40',$$
$$(\log \csc)^{-1}\, 0.1433 = 46°.01 - .04 = 45°.97 = 45°\, 58'.$$

## § 9. HYPERBOLIC FUNCTIONS.

*a.* The **hyperbolic functions** are certain functions which bear relations to the equilateral hyperbola similar to those borne by the circular functions to the circle; and they may often be usefully employed both in computation and in analysis. They are named the hyperbolic sine, cosine, tangent, cotangent, secant, and cosecant; and are variously denoted by different writers. They are here represented by the symbols: Sh, Ch, Th, Cth, Sch, Csch. They may be defined by the following formulas, in which

$$\mathsf{G} = \text{the exponential base}$$
$$= 1 + \frac{1}{1} + \frac{1}{1.2} + \frac{1}{1.2.3} + \frac{1}{1.2.3.4} + \cdots.$$
$$= 2.7182818285\ldots : -$$

$$\operatorname{Sh} x = \tfrac{1}{2}(\mathsf{G}^x - \mathsf{G}^{-x}), \qquad \operatorname{Ch} x = \tfrac{1}{2}(\mathsf{G}^x + \mathsf{G}^{-x}), \qquad \operatorname{Th} x = \frac{\operatorname{Sh} x}{\operatorname{Ch} x},$$

$$\operatorname{Cth} x = \frac{1}{\operatorname{Th} x}, \qquad \operatorname{Sch} x = \frac{1}{\operatorname{Ch} x}, \qquad \operatorname{Csch} x = \frac{1}{\operatorname{Sh} x}.$$

41

They bear to the circular functions the relations expressed by the following formulas, in which $i = \sqrt{-1}$ : —

$$\text{Sh } x = \frac{\sin xi}{i}, \qquad\qquad \sin x = \frac{\text{Sh } xi}{i},$$

$$\text{Ch } x = \cos xi, \qquad\qquad \cos x = \text{Ch } xi,$$

$$\text{Th } x = \frac{\tan xi}{i}, \qquad\qquad \tan x = \frac{\text{Th } xi}{i},$$

$$\text{Cth } x = i \text{ ctn } xi, \qquad\qquad \text{ctn } x = i \text{ Cth } xi,$$

$$\text{Sch } x = \sec xi, \qquad\qquad \sec x = \text{Sch } xi,$$

$$\text{Csch } x = i \text{ csc } xi, \qquad\qquad \text{csc } x = i \text{ Csch } xi.$$

Again, if $\phi$ is so taken that

$$x = \text{nat log tan } (45^\circ + \tfrac{1}{2}\phi),$$

$$\text{Sh } x = \tan \phi, \qquad \text{Ch } x = \sec \phi, \qquad \text{Csch } x = \text{ctn } \phi,$$
$$\text{Th } x = \sin \phi, \qquad \text{Sch } x = \cos \phi, \qquad \text{Cth } x = \text{csc } \phi.$$

The value of $\phi$ determined by this formula has been called by some writers the **Gudermannian** of $x$, and denoted by the symbol : gd $x$.

b. From $x = 0.00$ to $x = 1.00$, the function tabulated is gd $x$ in degrees, at intervals of 0.01 in the value of $x$. The hyperbolic functions of $x$ are then readily found, by the aid of the formulas last given, from the tables of circular functions. Beginning with $x = 1.00$, log Sh $x$, log Ch $x$, and log Th $x$ are tabulated, at intervals of 0.01 in the value of $x$, up to $x = 3.00$, the characteristic of each logarithm being placed at the head of its column ; then at intervals of 0.1 up to $x = 6.0$; and lastly at intervals of 1 up to $x = 10.0$. The printed differences are to be used, as in other tables, each through *half the interval* before and after the line on which it stands.

c. If $x > 10$, log Th $x = 0.0000$, while log Sh $x$ and log Ch $x$ may be found by the formula and table given at the lower right-hand corner of p. 21. The quantity $\mu$ is the *modulus* of the denary system of logarithms; that is, it is the denary logarithm of the exponential base. The values of $n\mu$ being given for all integral values of $n$ from 1 to 10, any product $x\mu$ is readily found, by adding together the products of $\mu$ by the successive figures of $x$. Only four decimal-figures should be retained in the result.

d. The functions log Cth $x$, log Sch $x$, and log Csch $x$ are the *arithmetical complements* of log Th $x$, log Ch $x$, and log Sh $x$, respectively.

e. The table may be used both directly and inversely. Examples : —

$$
\begin{aligned}
&\text{log Sh } 0.5378 && = \text{log tan } 20^\circ.427 = 9.7513, \\
&\text{log Ch } 1.5280 && = 0.3825, \\
&\text{log Cth } 1.8240 && = 0.0226, \\
&\text{log Sh } 12.5913 && = \text{log Ch } 12.5913 = 5.1673, \\
&(\text{log Ch})^{-1}\, 1.6000 && = 4.377, \\
&(\text{log Sh})^{-1}\, 5.0000 && = 12.206, \\
&(\text{log Th})^{-1}\, 9.9012 && = 1.089, \\
&(\text{log Sch})^{-1}\, 9.5873 && = 1.004, \\
&(\text{log Csch})^{-1}\, 0.3924 && = \text{gd}^{-1}\, (\text{log tan})^{-1}\, 9.6076 \\
&\qquad\qquad = \text{gd}^{-1}\, 22^\circ.06 && = 0.395.
\end{aligned}
$$

## § 10. NATURAL LOGARITHMS.

*a.* The **natural** system of logarithms is that which is founded on the **exponential base** (see § 9). This number is defined as *the limiting value* to which the expression

$$(1 + \epsilon)^{\frac{1}{\epsilon}} = \sqrt[\epsilon]{(1 + \epsilon)}$$

approaches, as $\epsilon$ approaches 0. It is most frequently denoted by the letter $e$; but, as being one of the few peculiar constants of analysis, it is here represented by the symbol $\ominus$, which may be read "**base.**"

The following formulas are proved in treatises on the Differential Calculus : —

$$\ominus = 1 + \frac{1}{1} + \frac{1}{1.2} + \frac{1}{1.2.3} + \frac{1}{1.2.3.4} + \cdots,$$

$$\ominus^x = 1 + \frac{x}{1} + \frac{x^2}{1.2} + \frac{x^3}{1.2.3} + \frac{x^4}{1.2.3.4} + \cdots,$$

$$\text{nat log } (1 + x) = x - \frac{x^2}{2} + \frac{x^3}{3} - \frac{x^4}{4} + \frac{x^5}{5} - \cdots ;$$

the second formula being applicable to *all values* of $x$, but the last only when $x$ is *numerically less than* 1. If $x$ is *very small*, then *approximately*

$$\ominus^x = 1 + x, \qquad \text{nat log } (1 + x) = x, \qquad \text{nat log } (1 - x) = -x.$$

We also have, in the natural system,

$$\log (a + h) = \log a + \log \left(1 + \frac{h}{a}\right) = \log a + \frac{h}{a} - \frac{h^2}{2a^2} + \frac{h^3}{3a^3} - \cdots,$$

provided $h$ is numerically less than $a$.

The *rate of difference* of $\ominus^x$, for $\Delta x = 1$, is always $\ominus^x$, and that of nat log $x$ is $\frac{1}{x}$.

*b.* The **numerical value** of $\ominus$ or of any power of $\ominus$ can be computed, to *any assigned number of decimal-places*, by using a sufficient number of terms of the first two series given above. Thus, to find $\ominus$ to four decimal-places, we proceed as follows, observing that, if any term be divided by its number in the series, the next following term is obtained : —

$$
\begin{aligned}
1) &\quad 1.00000 \\
2) &\quad 1.00000 \\
3) &\quad 0.50000 \\
4) &\quad 0.16667 \\
5) &\quad 0.04167 \\
6) &\quad 0.00833 \\
7) &\quad 0.00139 \\
8) &\quad 0.00020 \\
&\quad 0.00002 \\
\hline
\ominus = &\quad 2.7183 \ldots.
\end{aligned}
$$

43

# Explanation of the Tables.

*c.* The **modulus** of any system of logarithms is the logarithm of $\epsilon$ in that system. If $m$ is the modulus of a system of which $a$ is the base, then

$$a^m = \epsilon, \qquad \epsilon^{m-1} = a.$$

The modulus of the *natural system itself* is 1. The values of the modulus of the *denary system* and of the reciprocal of that modulus are

$$\mu \;\; = \text{den log } \epsilon = 0.4342944819\ldots,$$
$$\mu^{-1} = \text{nat log } 10 = 2.3025850930\ldots.$$

By the rule for converting logarithms from one system to another, the logarithm of a number in any system may be found by *multiplying the modulus of that system into the natural logarithm of the same number.* Thus,

$$\text{den log } x = \mu \text{ nat log } x,$$
$$\text{nat log } x = \mu^{-1} \text{ den log } x.$$

By the aid of these formulas, the table at the bottom of p. 21 may be used to find the natural logarithm of any number, or the denary logarithm of any power of the exponential base, or to find a number from its natural logarithm. For example : —

$$\text{nat log } 72.5 \;\; = 1.8603 \times \mu^{-1} = 4.2835,$$
$$\text{nat log } 1.0074 = 0.0032 \times \mu^{-1} = 0.0074,$$
$$\text{den log } \epsilon^{\frac{1}{7}} \;\; = \tfrac{1}{7}\mu \;\;\;\;\;\;\;\;\; = 0.0620,$$

$$(\text{nat log})^{-1}\, 10.2108 = (\text{den log})^{-1}\,(10.2108 \times \mu)$$
$$= (\text{den log})^{-1}\, 4.4345 = 27104.$$

*d.* The natural system is so called, because, in the higher mathematics, it is convenient to regard all other systems as founded upon this. It is named by some writers **hyperbolic**, and by others **Neperian**. But in fact, it is not the system of Napier ; nor has it any other relation to the hyperbola than that which belongs to logarithms in general.

*e.* We may make the following statement of the relation of logarithms and of the hyperbolic functions to the hyperbola, using the notation of Analytic Geometry : —

Let $xy = 1$ be the equation of an hyperbola referred to its asymptotes. It can be proved by the Integral Calculus that the *area*, contained between the curve and the axis of $x$, and between two ordinates of which one is drawn to the vertex of the curve, is measured by $\log x$ in the system of which the modulus is $\sin \omega$. Thus, *the logarithms belonging to any system may be represented by the areas of an appropriate hyperbola.* The natural system corresponds to the equilateral hyperbola, for which $\sin \omega = 1$.

Again, if $u$ denotes twice the area of the *sector* of the hyperbola $x^2 - y^2 = 1$, contained between the axis of $x$ and a radius vector from the centre, then

$$x = \text{Ch } u, \qquad y = \text{Sh } u\ ;$$

just as, in the circle $x^2 + y^2 = 1$, with a similar meaning of $u$,

$$x = \cos u, \qquad y = \sin u.$$